HUMANKIND

HUMANKIND

A Hopeful History

Rutger Bregman

Translated from the Dutch by
Elizabeth Manton and Erica Moore

Little, Brown and Company
New York Boston London

Little, Brown and Company
Hachette Book Group
1290 Avenue of the Americas, New York, NY 10104
littlebrown.com

First English-language edition: June 2020
Published simultaneously in the United Kingdom by Bloomsbury Publishing
Originally published September 2019 in the Netherlands as *De meeste mensen deugen* by De Correspondent (de correspondent.nl), a member-funded journalism platform for independent voices

Little, Brown and Company is a division of Hachette Book Group, Inc. The Little, Brown name and logo are trademarks of Hachette Book Group, Inc.

The publisher is not responsible for websites (or their content) that are not owned by the publisher.

The Hachette Speakers Bureau provides a wide range of authors for speaking events. To find out more, go to hachettespeakersbureau.com or call (866) 376-6591.

ISBN 978-0-316-41853-9 (hardcover) / 978-0-316-49881-4 (international paperback)
LCCN 2020930773

10 9 8 7 6 5 4

LSC-C

Printed in the United States of America

To my parents

CONTENTS

CONTENTS

'Man will become better when you show him what he is like.'

Anton Chekhov (1860–1904)

On the eve of the Second World War, the British Army Command found itself facing an existential threat. London was in grave danger. The city, according to a certain Winston Churchill, formed 'the greatest target in the world, a kind of tremendous fat cow, a valuable fat cow tied up to attract the beasts of prey'.[1]

The beast of prey was, of course, Adolf Hitler and his war machine. If the British population broke under the terror of his bombers, it would spell the end of the nation. 'Traffic will cease, the homeless will shriek for help, the city will be in pandemonium,' feared one British general.[2] Millions of civilians would succumb to the strain, and the army wouldn't even get around to fighting because it would have its hands full with the hysterical masses. Churchill predicted that at least three to four million Londoners would flee the city.

Anyone wanting to read up on all the evils to be unleashed needed only one book: *Psychologie des foules* – 'The Psychology of the Masses' – by one of the most influential scholars of his day, the Frenchman Gustave Le Bon. Hitler read the book cover to cover. So did Mussolini, Stalin, Churchill and Roosevelt.

Le Bon's book gives a play by play of how people respond to crisis. Almost instantaneously, he writes, 'man descends several rungs in the ladder of civilization'.[3] Panic and violence erupt, and we humans reveal our true nature.

On 19 October 1939, Hitler briefed his generals on the German plan of attack. 'The ruthless employment of the *Luftwaffe* against the heart of the British will-to-resist,' he said, 'can and will follow at the given moment.'[4]

In Britain, everyone felt the clock ticking. A last-ditch plan to dig a network of underground shelters in London was considered, but ultimately scrapped over concerns that the populace, paralysed by fear, would never re-emerge. At the last moment, a few psychiatric field hospitals were thrown up outside the city to tend to the first wave of victims.

And then it began.

On 7 September 1940, 348 German bomber planes crossed the Channel. The fine weather had drawn many Londoners outdoors, so when the sirens sounded at 4:43 p.m. all eyes went to the sky.

That September day would go down in history as Black Saturday, and what followed as 'the Blitz'. Over the next nine months, more than 80,000 bombs would be dropped on London alone. Entire neighbourhoods were wiped out. A million buildings in the capital were damaged or destroyed, and more than 40,000 people in the UK lost their lives.

So how did the British react? What happened when the country was bombed for months on end? Did people get hysterical? Did they behave like brutes?

Let me start with the eyewitness account of a Canadian psychiatrist.

In October 1940, Dr John MacCurdy drove through southeast London to visit a poor neighbourhood that had been particularly hard hit. All that remained was a patchwork of craters and crumbling buildings. If there was one place sure to be in the grip of pandemonium, this was it.

So what did the doctor find, moments after an air raid alarm? 'Small boys continued to play all over the pavements, shoppers went on haggling, a policeman directed traffic in majestic boredom and the bicyclists defied death and the traffic laws. No one, so far as I could see, even looked into the sky.'[5]

In fact, if there's one thing that all accounts of the Blitz have in common it's their description of the strange serenity that settled over London in those months. An American journalist interviewing a British couple in their kitchen noted how they sipped tea even as the windows rattled in their frames. Weren't they afraid?, the journalist wanted to know. 'Oh no,' was the answer. 'If we were, what good would it do us?'[6]

Evidently, Hitler had forgotten to account for one thing: the quintessential British character. The stiff upper lip. The wry humour, as expressed by shop owners who posted signs in front of their wrecked premises announcing: MORE OPEN THAN USUAL. Or the pub proprietor who in the midst of devastation advertised: OUR WINDOWS ARE GONE, BUT OUR SPIRITS ARE EXCELLENT. COME IN AND TRY THEM.[7]

The British endured the German air raids much as they would a delayed train. Irritating, to be sure, but tolerable on the whole. Train services, as it happens, also continued during the Blitz, and Hitler's tactics scarcely left a dent in the domestic economy. More detrimental to the British war machine was Easter Monday in April 1941, when everybody had the day off.[8]

Within weeks after the Germans launched their bombing campaign, updates were being reported much like the weather: 'Very blitzy tonight.'[9] According to an American observer, 'the English get bored so much more quickly than they get anything else, and nobody is taking cover much any longer'.[10]

And the mental devastation, then? What about the millions of traumatised victims the experts had warned about? Oddly enough, they were nowhere to be found. To be sure, there was sadness and fury; there was terrible grief at the loved ones lost. But the psychiatric wards remained empty. Not only that, public mental health actually improved. Alcoholism tailed off. There were fewer suicides than in peacetime. After the war ended, many British would yearn for the days of the Blitz, when everybody helped each other out and no one cared about your politics, or whether you were rich or poor.[11]

'British society became in many ways strengthened by the Blitz,' a British historian later wrote. 'The effect on Hitler was disillusioning.'[12]

When put to the test, the theories set forth by celebrated crowd psychologist Gustave Le Bon could hardly have been further off the mark. Crisis brought out not the worst, but the *best* in people. If anything, the British moved up a few rungs on the ladder of civilisation. 'The courage, humor, and kindliness of ordinary people,' an American journalist confided in her diary, 'continue to be astonishing under conditions that possess many of the features of a nightmare.'[13]

These unexpected impacts of the German bombings sparked a debate on strategy in Britain. As the Royal Air Force prepared to deploy its own fleet of bombers against the enemy, the question was how to do so most effectively.

Curiously, given the evidence, the country's military experts still espoused the idea that a nation's morale could be broken. By bombs. True, it hadn't worked on the British, the reasoning went, but they were a special case. No other people on the planet could match their levelheadedness and fortitude. Certainly not the Germans, whose fundamental 'lack of moral fibre' meant they would 'not stand a quarter of the bombing' the British endured.[14]

Among those who endorsed this view was Churchill's close friend Frederick Lindemann, also known as Lord Cherwell. A rare photograph of him shows a tall man with a cane, wearing a bowler hat and an icy expression.[15] In the fierce debate over air strategy, Lindemann remained adamant: bombing *works*. Like Gustave Le Bon, he took a dim view of the masses, writing them off as cowardly and easily panicked.

To prove his point, Lindemann dispatched a team of psychiatrists to Birmingham and Hull, two cities where the German bombings had taken an especially heavy toll. They interviewed hundreds of men, women and children who had lost their homes during the Blitz, inquiring about the smallest details – 'down to the number of pints drunk and aspirins bought in the chemists'.[16]

The team reported back to Lindemann a few months later. The conclusion, printed in large letters on the title page, was this:

THERE IS NO EVIDENCE OF BREAKDOWN OF MORALE.[17]

So what did Frederick Lindemann do with this unequivocal finding? He ignored it. Lindemann had already decided that strategic bombing was a sure bet, and mere facts were not about to change his mind.

And so the memo he sent to Churchill said something altogether different:

Investigation seems to show that having one's house demolished is most dangerous to morale. People seem to mind it more than having their friends or even relatives killed. At Hull, signs of strain were evident though only one-tenth of the homes were demolished. On the above figures, we can do as much harm to each of the 58 principal

German towns. There seems little doubt that this would break the spirit of the German people.[18]

Thus ended the debate over the efficacy of bombing. The whole episode had, as one historian later described it, the 'perceptible smell of a witch hunt'.[19] Conscientious scientists who opposed the tactic of targeting German civilians were denounced as cowards, even traitors.

The bomb-mongers, meanwhile, felt the enemy needed to be dealt an even harsher blow. Churchill gave the signal and all hell broke loose over Germany. When the bombing finally ended, the casualties numbered ten times higher than after the Blitz. On one night in Dresden, more men, women and children were killed than in London during the whole war. More than half of Germany's towns and cities were destroyed. The country had become one big heap of smouldering rubble.

All the while, only a small contingent of the Allied air force was actually striking strategic targets such as factories and bridges. Right up through the final months, Churchill maintained that the surest way to win the war was by dropping bombs on civilians to break national morale. In January 1944, a Royal Air Force memo gratifyingly affirmed this view: 'The more we bomb, the more satisfactory the effect.'

The prime minister underlined these words using his famous red pen.[20]

So did the bombings have the intended effect?

Let me again start with an eyewitness account from a respected psychiatrist. Between May and July 1945, Dr Friedrich Panse interviewed almost a hundred Germans whose homes had been destroyed. 'Afterward,' said one, 'I was really full of vim and lit up a cigar.' The general mood following a raid, said another, was euphoric, 'like after a war that has been won.'[21]

There was no sign of mass hysteria. On the contrary, in places that had just been hit, inhabitants felt relief. 'Neighbours were wonderfully helpful,' Panse recorded. 'Considering the severity and duration of the mental strain, the general attitude was remarkably steady and restrained.'[22]

Reports by the *Sicherheitsdienst*, which kept close tabs on the German population, convey a similar picture. After the raids, people helped each other out. They pulled victims from the rubble, they extinguished fires. Members of the Hitler Youth rushed around tending to the homeless and the injured. A grocer jokingly hung up a sign in front of his shop: DISASTER BUTTER SOLD HERE![23]

(Okay, the British humour was better.)

Shortly after the German surrender in May 1945, a team of Allied economists visited the defeated nation, tasked by the US Department of Defense to study the effects of the bombing. Most of all, the Americans wanted to know if this tactic was a good way to win wars.

The scientists' findings were stark: the civilian bombings had been a fiasco. In fact, they appeared to have strengthened the German wartime economy, thereby prolonging the war. Between 1940 and 1944, they found that German tank production had multiplied by a factor of nine, and of fighter jets by a factor of *fourteen*.

A team of British economists reached the same conclusion.[24] In the twenty-one devastated towns and cities they investigated, production had increased faster than in a control group of fourteen cities that had not been bombed. 'We were beginning to see,' confessed one of the American economists, 'that we were encountering one of the greatest, perhaps the greatest miscalculation of the war.'[25]

What fascinates me most about this whole sorry affair is that the main actors all fell into the same trap.

Hitler and Churchill, Roosevelt and Lindemann – all of them signed on to psychologist Gustave Le Bon's claim that our state of civilisation is no more than skin deep. They were certain that air raids would blow this fragile covering to bits. But the more they bombed, the *thicker* it got. Seems it wasn't a thin membrane at all, but a callus.

Military experts, unfortunately, were slow to catch on. Twenty-five years later, US forces would drop three times as much firepower on Vietnam as they dropped in the entire Second World War.[26] This time it failed on an even grander scale. Even when the evidence is right in front of us, somehow we still manage to deny it. To this day, many remain convinced that the resilience the British people showed during the Blitz can be chalked up to a quality that is singularly British.

But it's not singularly British. It's universally human.

A New Realism

I

This is a book about a radical idea.

An idea that's long been known to make rulers nervous. An idea denied by religions and ideologies, ignored by the news media and erased from the annals of world history.

At the same time, it's an idea that's legitimised by virtually every branch of science. One that's corroborated by evolution and confirmed by everyday life. An idea so intrinsic to human nature that it goes unnoticed and gets overlooked.

If only we had the courage to take it more seriously, it's an idea that might just start a revolution. Turn society on its head. Because once you grasp what it really means, it's nothing less than a mind-bending drug that ensures you'll never look at the world the same again

So what is this radical idea?

That most people, deep down, are pretty decent.

I don't know anyone who explains this idea better than Tom Postmes, professor of social psychology at the University of Groningen in the Netherlands. For years, he's been asking students the same question.

Imagine an airplane makes an emergency landing and breaks into three parts. As the cabin fills with smoke, everybody inside realises: We've got to get out of here. What happens?

On Planet A, the passengers turn to their neighbours to ask if they're okay. Those needing assistance are helped out of the plane first. People are willing to give their lives, even for perfect strangers.

On Planet B, everyone's left to fend for themselves. Panic breaks out. There's lots of pushing and shoving. Children, the elderly, and people with disabilities get trampled underfoot.

Now the question: Which planet do we live on?

'I would estimate about 97 per cent of people think we live on Planet B,' says Professor Postmes. 'The truth is, in almost every case, we live on Planet A.'[1]

Doesn't matter who you ask. Left wing or right, rich or poor, uneducated or well read – all make the same error of judgement. 'They don't know. Not freshman or juniors or grad students, not professionals in most cases, not even emergency responders,' Postmes laments. 'And it's not for a lack of research. We've had this information available to us since World War II.'

Even history's most momentous disasters have played out on Planet A. Take the sinking of the *Titanic*. If you saw the movie, you probably think everybody was blinded by panic (except the string quartet). In fact, the evacuation was quite orderly. One eyewitness recalled that 'there was no indication of panic or hysteria, no cries of fear, and no running to and fro'.[2]

Or take the September 11 2001 terrorist attacks. As the Twin Towers burned, thousands of people descended the stairs calmly, even though they knew their lives were in danger. They stepped aside for firefighters and the injured. 'And people would actually say: "No, no, you first," one survivor later reported. 'I couldn't believe it, that at this point people would actually say "No, no, please take my place." It was uncanny.'[3]

There is a persistent myth that by their very nature humans are selfish, aggressive and quick to panic. It's what Dutch biologist Frans de Waal likes to call *veneer theory*: the notion that civilisation is nothing more than a thin veneer that will crack at the merest provocation.[4] In actuality, the opposite is true. It's when crisis hits – when the bombs fall or the floodwaters rise – that we humans become our best selves.

On 29 August 2005, Hurricane Katrina tore over New Orleans. The levees and flood walls that were supposed to protect the city failed. In the wake of the storm, 80 per cent of area homes flooded and at least 1,836 people lost their lives. It was one of the most devastating natural disasters in US history.

That whole week newspapers were filled with accounts of rapes and shootings across New Orleans. There were terrifying reports of roving gangs, lootings and of a sniper taking aim at rescue helicopters. Inside the Superdome, which served as the city's largest storm shelter, some 25,000 people were packed in together, with no electricity and no water. Two infants' throats had been slit, journalists reported, and a seven-year-old had been raped and murdered.[5]

The chief of police said the city was slipping into anarchy, and the governor of Louisiana feared the same. 'What angers me the most,' she said, 'is that disasters like this often bring out the worst in people.'[6]

This conclusion went viral. In the British newspaper the *Guardian*, acclaimed historian Timothy Garton Ash articulated what so many were thinking: 'Remove the elementary staples of organised, civilised life – food, shelter, drinkable water, minimal personal security – and we go back within hours to a Hobbesian state of nature, a war of all against all. [...] A few become temporary angels, most revert to being apes.'

There it was again, in all its glory: veneer theory. New Orleans, according to Garton Ash, had opened a small hole in 'the thin crust we lay across the seething magma of nature, including human nature'.[7]

It wasn't until months later, when the journalists cleared out, the floodwaters drained away and the columnists moved on to their next opinion, that researchers uncovered what had really happened in New Orleans.

What sounded like gunfire had actually been a popping relief valve on a gas tank. In the Superdome, six people had died: four of natural causes, one from an overdose and one by suicide. The police chief was forced to concede that he couldn't point to a single officially reported rape or murder. True, there had been looting, but mostly by groups that had teamed up to survive, in some cases even banding with police.[8]

Researchers from the Disaster Research Center at the University of Delaware concluded that 'the overwhelming majority of the emergent activity was prosocial in nature'.[9] A veritable armada of boats from as far away as Texas came to save people from the rising waters. Hundreds of civilians formed rescue squads, like the self-styled Robin Hood Looters – a group of eleven friends who went around looking for food, clothing and medicine and then handing it out to those in need.[10]

Katrina, in short, didn't see New Orleans overrun with self-interest and anarchy. Rather, the city was inundated with courage and charity.

The hurricane confirmed the science on how human beings respond to disasters. Contrary to what we normally see in the movies, the Disaster Research Center at the University of

Delaware has established that in nearly seven hundred field studies since 1963, there's never total mayhem. It's never every man for himself. Crime – murder, burglary, rape – usually drops. People don't go into shock, they stay calm and spring into action. 'Whatever the extent of the looting,' a disaster researcher points out, 'it always pales in significance to the wide-spread altruism that leads to free and massive giving and sharing of goods and services.'[11]

Catastrophes bring out the best in people. I know of no other sociological finding that's backed by so much solid evidence that's so blithely ignored. The picture we're fed by the media is consistently the opposite of what happens when disaster strikes.

Meanwhile, back in New Orleans, all those persistent rumours were costing lives.

Unwilling to venture into the city unprotected, emergency responders were slow to mobilise. The National Guard was called in, and at the height of the operation some 72,000 troops were in place. 'These troops know how to shoot and kill,' said the governor, 'and I expect they will.'[12]

And so they did. On Danziger Bridge on the city's east side, police opened fire on six innocent, unarmed black residents, killing a seventeen-year-old boy and a mentally disabled man of forty (five of the officers involved were later sentenced to lengthy prison terms).[13]

True, the disaster in New Orleans was an extreme case. But the dynamic during disasters is almost always the same: adversity strikes and there's a wave of spontaneous cooperation in response, then the authorities panic and unleash a second disaster.

'My own impression,' writes Rebecca Solnit, whose book *A Paradise Built in Hell* (2009) gives a masterful account of Katrina's aftermath, 'is that elite panic comes from powerful people who see all humanity in their own image.'[14] Dictators and despots, governors and generals – they all too often resort to brute force to prevent scenarios that exist only in their own heads, on the assumption that the average Joe is ruled by self-interest, just like them.

2

In the summer of 1999, at a small school in the Belgian town of Bornem, nine children came down with a mysterious illness. They'd come to school that morning with no symptoms; after lunch they were all ill. Headaches. Vomiting. Palpitations. Casting about for an explanation, the only thing the teachers could think of was the Coca-Cola the nine had drunk during break.

It didn't take long for journalists to get wind of the story. Over at Coca-Cola headquarters, the phones started ringing. That same evening the company issued a press release stating that millions of bottles were being recalled from Belgian store shelves. 'We are searching frantically and hope to have a definitive answer in the next few days,' said a spokeswoman.[15]

But it was too late. The symptoms had spread through Belgium and jumped the border into France. Pale, limp kids were being rushed off in ambulances. Within days, suspicion had spread to all Coca-Cola products. Fanta, Sprite, Nestea, Aquarius . . . they all seemed a danger to children. The 'Coca-Cola Incident' was one of the worst financial blows in the company's 107-year history, forcing it to recall seventeen million cases of soft drinks in Belgium and destroy its warehoused stock.[16] In the end, the cost was more than 200 million dollars.[17]

Then something odd happened. A few weeks later, the toxicologists issued their lab report. What had they found after running their tests on the cans of Coke? Nothing. No pesticides.

No pathogens. No toxic metals. Nada. And their tests on the blood and urine samples from hundreds of patients? Zilch. The scientists were unable to find a single chemical cause for the severe symptoms which by that time had been documented in more than a thousand boys and girls.

'Those kids really were sick, there's no doubt about that,' said one of the researchers. 'But not from drinking a Coke.'[18]

The Coca-Cola incident speaks to an age-old philosophical question.

What is truth?

Some things are true whether you believe in them or not. Water boils at 100°C. Smoking kills. President Kennedy was assassinated in Dallas on 22 November 1963.

Other things have the potential to be true, if we believe in them. Our belief becomes what sociologists dub a *self-fulfilling prophecy*: if you predict a bank will go bust and that convinces lots of people to close their accounts, then, sure enough, the bank will go bust.

Or take the placebo effect. If your doctor gives you a fake pill and says it will cure what ails you, chances are you *will* feel better. The more dramatic the placebo, the bigger that chance. Injection, on the whole, is more effective than pills, and in the old days even bloodletting could do the trick – not because medieval medicine was so advanced, but because people felt a procedure that drastic was bound to have an impact.

And the ultimate placebo? Surgery! Don a white coat, administer an anaesthetic, and then kick back and pour yourself a cup of coffee. When the patient revives tell them the operation was a success. A broad review carried out by the *British Medical Journal* comparing actual surgical procedures with sham surgery (for conditions like back pain and heartburn) revealed that placebos also helped in three-quarters of all cases, and in half were just as effective as the real thing.[19]

But it also works the other way around.

Take a fake pill thinking it will make you sick, and chances are it will. Warn your patients a drug has serious side effects, and it probably will. For obvious reasons, the *nocebo* effect, as it's called, hasn't been widely tested, given the touchy ethics of convincing healthy people they're ill. Nevertheless, all the evidence suggests nocebos can be very powerful.

That's also what Belgian health officials concluded in the summer of 1999. Possibly there really was something wrong with one or two of the Cokes those kids in Bornem drank. Who's to say? But beyond that, the scientists were unequivocal: the hundreds of other children across the country had been infected with a 'mass psychogenic illness'. In plain English: they imagined it.

Which is not to say the victims were pretending. More than a thousand Belgian kids were genuinely nauseated, feverish and dizzy. If you believe something enough, it can become real. If there's one lesson to be drawn from the nocebo effect, it's that ideas are never *merely* ideas. We are what we believe. We find what we go looking for. And what we predict, comes to pass.

Maybe you see where I'm going with this: our grim view of humanity is also a nocebo.

If we *believe* most people can't be trusted, that's how we'll treat each other, to everyone's detriment. Few ideas have as much power to shape the world as our view of other people. Because ultimately, you get what you expect to get. If we want to tackle the greatest challenges of our times – from the climate crisis to our growing distrust of one another – then I think the place we need to start is our view of human nature.

To be clear: this book is not a sermon on the fundamental goodness of people. Obviously, we're not angels. We're complex creatures, with a good side and a not-so-good side. The question is which side we turn to.

My argument is simply this: that we — by nature, as children, on an uninhabited island, when war breaks out, when crisis hits — have a powerful preference for our good side. I will present the considerable scientific evidence showing just how realistic a more positive view of human nature is. At the same time, I'm convinced it could be more of a reality if we'd start to believe it.

Floating around the Internet is a parable of unknown origin. It contains what I believe is a simple but profound truth:

> An old man says to his grandson: 'There's a fight going on inside me. It's a terrible fight between two wolves. One is evil — angry, greedy, jealous, arrogant, and cowardly. The other is good — peaceful, loving, modest, generous, honest, and trustworthy. These two wolves are also fighting within you, and inside every other person too.'
> After a moment, the boy asks, 'Which wolf will win?'
> The old man smiles.
> 'The one you feed.'

3

Over the last few years, whenever I told people about this book I've been working on, I was met with raised eyebrows. Expressions of disbelief. A German publisher flatly turned down my book proposal. Germans, she said, don't believe in humanity's innate goodness. A member of the Parisian intelligentsia assured me that the French need government's firm hand. And when I toured the United States after the 2016 presidential election, everyone, everywhere, asked me if my head was screwed on straight.

Most people are decent? Had I ever turned on a television?

Not so long ago, a study by two American psychologists proved once again how stubbornly people can cling to the idea of our own selfish nature. The researchers presented test subjects with several situations featuring other people doing apparently nice things. So what did they find? Basically, that we are trained to see selfishness everywhere.

See someone helping an elderly person cross the street?

What a show-off.

See someone offering money to a homeless person?

Must want to feel better about herself.

Even after the researchers presented their subjects with hard data about strangers returning lost wallets, or the fact that the vast majority of the population doesn't cheat or steal, most subjects did not view humanity in a more positive light. 'Instead,' write the psychologists, 'they decide that seemingly selfless behaviors must be selfish after all.'[20]

Cynicism is a theory of everything. The cynic is always right.

Now, you may be thinking: wait a second, that's not how I was raised. Where I come from we trusted each other, helped each other and left our doors unlocked. And you're right, from up close, it's easy to assume people are decent. People like our families and friends, our neighbours and our co-workers.

But when we zoom out to the rest of humanity, suspicion quickly takes over. Take the World Values Survey, a huge poll conducted since the 1980s by a network of social scientists in almost a hundred countries. One standard question is: 'Generally speaking, would you say that most people can be trusted or that you need to be very careful in dealing with people?'

The results are pretty disheartening. In nearly every country most people think most other people can't be trusted. Even in established democracies like France, Germany, Great Britain and the United States, the majority of the population shares this poor view of their fellow human beings.[21]

The question that has long fascinated me is *why* we take such a negative view of humanity. When our instinct is to trust those in our immediate communities, why does our attitude change when applied to people as a whole? Why do so many laws and regulations, so many companies and institutions start with the assumption that people can't be trusted? Why, when the science consistently tells us we live on Planet A, do we persist in believing we're on Planet B?

Is it a lack of education? Hardly. In this book I will introduce dozens of intellectuals who are staunch believers in our immorality. Political conviction? No again. Quite a few religions take it as a tenet of faith that humans are mired in sin. Many a capitalist presumes we're all motivated by self-interest. Lots of environmentalists see humans as a destructive plague upon the earth. Thousands of opinions; one take on human nature.

This got me wondering. Why do we imagine humans are bad? What made us start believing in the wicked nature of our kind?

Imagine for a moment that a new drug comes on the market. It's super-addictive, and in no time everyone's hooked. Scientists investigate and soon conclude that the drug causes, I quote, 'a misperception of risk, anxiety, lower mood levels, learned helplessness, contempt and hostility towards others, [and] desensitization'.[22]

Would we use this drug? Would our kids be allowed to try it? Would government legalise it? To all of the above: yes. Because what I'm talking about is already one of the biggest addictions of our times. A drug we use daily, that's heavily subsidised and is distributed to our children on a massive scale.

That drug is the news.

I was raised to believe that the news is good for your development. That as an engaged citizen it's your duty to read the paper and watch the evening news. That the more we follow the news, the better informed we are and the healthier our democracy. This is still the story many parents tell their kids, but scientists are reaching very different conclusions. The news, according to dozens of studies, is a mental health hazard.[23]

First to open up this field of research, back in the 1990s, was George Gerbner (1919–2005). He also coined a term to describe the phenomenon he found: *mean world syndrome*, whose clinical symptoms are cynicism, misanthropy and pessimism. People who follow the news are more likely to agree with statements such as 'Most people care only about themselves.' They more often believe that we as individuals are helpless to better the world. They are more likely to be stressed and depressed.

A few years ago, people in thirty different countries were asked a simple question: 'Overall, do you think the world is getting better, staying the same, or getting worse?' In every country, from Russia to Canada, from Mexico to Hungary, the vast majority of people answered that things are getting *worse*.[24] The reality is exactly the opposite. Over the last several decades, extreme poverty, victims of war, child mortality, crime, famine, child labour, deaths in natural disasters and the number of plane crashes have all plummeted. We're living in the richest, safest, healthiest era ever.

So why don't we realise this? It's simple. Because the news is about the exceptional, and the more exceptional an event is – be it a terrorist attack, violent uprising, or natural disaster – the bigger its newsworthiness. You'll never see a headline reading NUMBER OF PEOPLE LIVING IN EXTREME POVERTY

DOWN BY 137,000 SINCE YESTERDAY, even though it could accurately have been reported *every day over the last twenty-five years*.[25] Nor will you ever see a broadcast go live to a reporter on the ground who says, 'I'm standing here in the middle of nowhere, where today there's still no sign of war.'

A couple of years ago, a team of Dutch sociologists analysed how aeroplane crashes are reported in the media. Between 1991 and 2005, when the number of accidents consistently dropped, they found media attention for such accidents consistently grew. And as you might expect, people grew increasingly fearful to fly on these increasingly safe planes.[26]

In another study, a team of media researchers compiled a database of over four million news items on immigration, crime and terrorism in order to determine if there were any patterns. What they found is that in times when immigration or violence declines, newspapers give them *more* coverage. 'Hence,' they concluded, 'there seems to be none or even a negative relationship between news and reality.'[27]

Of course, by 'the news' I don't mean all journalism. Many forms of journalism help us better understand the world. But the news – by which I mean reporting on recent, incidental and sensational events – is most common. Eight in ten adults in western countries are daily news consumers. On average, we spend one hour a day getting our news fix. Added up over a lifetime, that's three years.[28]

Why are we humans so susceptible to the doom and gloom of the news? Two reasons. The first is what psychologists call *negativity bias*: we're more attuned to the bad than the good. Back in our hunting and gathering days, we were better off being frightened of a spider or a snake a hundred times too often than one time too few. Too much fear wouldn't kill you; too little surely would.

Second, we're also burdened with an *availability bias*. If we can easily recall examples of a given thing, we assume that thing is relatively common. The fact that we're bombarded daily with horrific stories about aircraft disasters, child snatchers and beheadings – which tend to lodge in the memory – completely skews our view of the world. As the Lebanese statistician Nassim Nicholas Taleb dryly notes, 'We are not rational enough to be exposed to the press'.[29]

In this digital age, the news we're being fed is only getting more extreme. In the old days, journalists didn't know much about their individual readers. They wrote for the masses. But the people behind Facebook, Twitter and Google know you well. They know what shocks and horrifies you, they know what makes you click. They know how to grab your attention and hold it so they can serve you the most lucrative helping of personalised ads.

This modern media frenzy is nothing less than an assault on the mundane. Because, let's be honest, the lives of most people are pretty predictable. Nice, but boring. So while we'd prefer having nice neighbours with boring lives (and thankfully most neighbours fit the bill), 'boring' won't make you sit up and take notice. 'Nice' doesn't sell ads. And so Silicon Valley keeps dishing us up ever more sensational clickbait, knowing full well, as a Swiss novelist once quipped, that 'News is to the mind what sugar is to the body.'[30]

A few years ago I resolved to make a change. No more watching the news or scrolling through my phone at breakfast. From now on, I would reach for a good book. About history. Psychology. Philosophy.

Pretty soon, however, I noticed something familiar. Most books are also about the exceptional. The biggest history bestsellers are invariably about catastrophes and adversity, tyranny and oppression. About war, war, and, to spice things up a little, war. And if, for once, there is no war, then we're in what historians call the *interbellum*: between wars.

In science, too, the view that humanity is bad has reigned for decades. Look up books on human nature and you'll find titles like *Demonic Males*, *The Selfish Gene* and *The Murderer Next Door*. Biologists long assumed the gloomiest theory of evolution, where even if an animal *appeared* to do something kind, it was framed as selfish. Familial affection? Nepotism! Monkey splits a banana? Exploited by a freeloader![31] As one American biologist mocked, 'What passes for co-operation turns out to be a mixture of opportunism and exploitation. [...] Scratch an "altruist" and watch a "hypocrite" bleed.'[32]

And in economics? Much the same. Economists defined our species as the *homo economicus*: always intent on personal gain, like selfish, calculating robots. Upon this notion of human nature, economists built a cathedral of theories and models that wound up informing reams of legislation.

Yet no one had researched whether *homo economicus* actually existed. That is, not until economist Joseph Henrich and his team took it up in 2000. Visiting fifteen communities in twelve countries on five continents, they tested farmers, nomads, and hunters and gatherers, all in search of this hominid that has guided economic theory for decades. To no avail. Each and every time, the results showed people were simply too decent. Too kind.[33]

After publishing this influential finding, Henrich continued his quest for the mythical being around which so many economists had spun their theories. Eventually he found him: *homo economicus* in the flesh. Although *homo* is not quite the right word. *Homo economicus*, it turns out, is not a human, but a chimpanzee. 'The canonical predictions of the *Homo economicus* model have proved remarkably successful in predicting chimpanzee behaviour in simple experiments,' Henrich noted dryly. 'So, all theoretical work was not wasted, it was just applied to the wrong species.'[34]

Less amusing is that this dim view of human nature has worked as a nocebo for decades now. In the 1990s, economics professor Robert Frank wondered how viewing humans as ultimately egotistical might affect his students. He gave them a range of assignments designed to gauge their generosity. The outcome? The longer they'd studied economics, the more selfish they'd become. 'We become what we teach,' Frank concluded.[35]

The doctrine that humans are innately selfish has a hallowed tradition in the western canon. Great thinkers like Thucydides, Augustine, Machiavelli, Hobbes, Luther, Calvin, Burke, Bentham, Nietzsche, Freud and America's Founding Fathers each had their own version of the veneer theory of civilisation. They all assumed we live on Planet B.

This cynical view was already circulating among the ancient Greeks. We read it in the writings of one of the first historians, Thucydides, when he describes a civil war that broke out on the Greek island of Corcyra in 427 BCE. 'With the ordinary conventions of civilized life thrown into confusion,' he wrote, 'human nature, always ready to offend even where laws exist, showed itself proudly in its true colours.'[36] That is to say, people behaved like beasts.

A negative outlook has also permeated Christianity from its early days. The Church Father Augustine (354–430) helped popularise the idea that humans are born sinful. 'No one is free from sin,' he wrote, 'not even an infant whose span of earthly life is but a single day.'[37]

This concept of original sin remained popular through the Reformation, when Protestants broke with the Roman Catholic Church. According to theologian and reformer John Calvin, 'our nature is not only destitute and empty of good, but so fertile and fruitful of every evil that it cannot be idle.' This belief was encoded in key Protestants texts like the Heidelberg Catechism (1563), which informs us that humans are 'totally unable to do any good and inclined to all evil'.

Weirdly, not only traditional Christianity but also the Enlightenment, which placed reason over faith, is rooted in a grim view of human nature. Orthodox faithful were convinced our kind is essentially depraved and the best we can do is apply a thin gloss of piety. Enlightenment philosophers also thought we were depraved, but prescribed a coating of reason to cover the rot.

When it comes to notions about human nature, the continuity throughout Western thought is striking. 'For this can be said of men in general: that they are ungrateful, fickle, hypocrites,' summed up the founder of political science, Niccolò Machiavelli. 'All men would be tyrants if they could,' agreed John Adams, founder of American democracy. 'We are descended from an endless series of generations of murderers,' diagnosed Sigmund Freud, founder of modern psychology.

In the nineteenth century Charles Darwin burst onto the scene with his theory of evolution, and it too was swiftly given the veneer treatment. The renowned scientist Thomas Henry Huxley (aka 'Darwin's Bulldog') preached that life is one great battle 'of man against man and of nation against nation'.[38] The philosopher Herbert Spencer sold hundreds of thousands of books on his assertion that we should fan the flames of this battle, since 'the whole effort of Nature is to get rid of [the poor] – to clear the world of them, and make room for better'.[39]

Strangest of all is that these thinkers were almost unanimously hailed as 'realists', while dissident thinkers were ridiculed for believing in human decency.[40] Emma Goldman, a feminist whose struggle for freedom and equality earned her a lifetime of slander and contempt, once wrote: 'Poor human nature, what horrible crimes have been committed in thy name! [...] The greater the mental charlatan, the more definite his insistence on the wickedness and weaknesses of human nature.'[41]

Only recently have scientists from an array of different fields come to the conclusion that our grim view of humanity is due for radical revision. This awareness is still so incipient that many of them don't realise they have company. As one prominent psychologist exclaimed when I told her about the new currents in biology: 'Oh God, so it's happening there as well?'[42]

4

Before I report on my quest for a new view of humankind, I want to share three warnings.

First, to stand up for human goodness is to stand up against a hydra – that mythological seven-headed monster that grew back two heads for every one Hercules lopped off. Cynicism works a lot like that. For every misanthropic argument you deflate, two more will pop up in its place. Veneer theory is a zombie that just keeps coming back.

Second, to stand up for human goodness is to take a stand against the powers that be. For the powerful, a hopeful view of human nature is downright threatening. Subversive. Seditious. It implies that we're not selfish beasts that need to be reined in, restrained and regulated. It implies that we need a different kind of leadership. A company with intrinsically motivated employees has no need of managers; a democracy with engaged citizens has no need of career politicians.

Third, to stand up for human goodness means weathering a storm of ridicule. You'll be called naive. Obtuse. Any weakness in your reasoning will be mercilessly exposed. Basically, it's easier to be a cynic. The pessimistic professor who preaches the doctrine of human depravity can predict anything he wants, for if his prophecies don't come true now, just wait: failure could always be just around the corner. Or else, his voice of reason has prevented the worst. The prophets of doom sound oh so profound, whatever they spout.

The reasons for hope, by contrast, are always provisional. Nothing has gone wrong – yet. You haven't been cheated – yet. An idealist can be right her whole life and still be dismissed as naive. This book is intended to change that. Because what seems unreasonable, unrealistic and impossible today can turn out to be inevitable tomorrow.

The time has come for a new view of human nature. It's time for a new realism. It's time for a new view of humankind.

2

The Real *Lord of the Flies*

I

When I started writing this book, I knew there was one story I would have to address.

The story takes place on a deserted island somewhere in the Pacific. A plane has just gone down. The only survivors are some British schoolboys, who can't believe their good fortune. It's as if they've just crash-landed in one of their adventure books. Nothing but beach, shells and water for miles. And better yet: no grown-ups.

On the very first day, the boys institute a democracy of sorts. One boy – Ralph – is elected to be the group's leader. Athletic, charismatic and handsome, he's the golden boy of the bunch. Ralph's game plan is simple: 1) Have fun. 2) Survive. 3) Make smoke signals for passing ships.

Number one is a success. The others? Not so much. Most of the boys are more interested in feasting and frolicking than in tending the fire. Jack, the redhead, develops a passion for hunting pigs and as time progresses he and his friends grow increasingly reckless. When a ship does finally pass in the distance, they've abandoned their post at the fire.

'You're breaking the rules!' Ralph accuses angrily.

Jack shrugs. 'Who cares?'

'The rules are the only thing we've got!'

When night falls, the boys are gripped by terror, fearful of the beast they believe is lurking on the island. In reality, the only beast is inside them. Before long, they've begun painting their faces. Casting off their clothes. And they develop overpowering urges – to pinch, to kick, to bite.

Of all the boys, only one manages to keep a cool head. Piggy, as the others call him because he's pudgier than the rest, has asthma, wears glasses and can't swim. Piggy is the voice of reason, to which nobody listens. 'What are we?' he wonders mournfully. 'Humans? Or animals? Or savages?'

Weeks pass. Then, one day, a British naval officer comes ashore. The island is now a smouldering wasteland. Three of the children, including Piggy, are dead. 'I should have thought,' the officer reproaches them, 'that a pack of British boys would have been able to put up a better show than that.' Ralph, the leader of the once proper and well-behaved band of boys, bursts into tears.

'Ralph wept for the end of innocence,' we read, and for 'the darkness of man's heart ...'

This story never happened. An English schoolmaster made it up in 1951. 'Wouldn't it be a good idea,' William Golding asked his wife one day, 'to write a story about some boys on an island, showing how they would really behave?'[1]

Golding's book *Lord of the Flies* would ultimately sell tens of millions of copies, be translated into more than thirty languages and be hailed as one of the classics of the twentieth century.

In hindsight, the secret to the book's success is clear. Golding had a masterful ability to portray the darkest depths of mankind. 'Even if we start with a clean slate,' he wrote in his first letter to his publisher, 'our nature compels us to make a muck of it.'[2] Or as he later put it, 'Man produces evil as a bee produces honey.'[3]

Of course, Golding had the *zeitgeist* of the 1960s on his side, when a new generation was questioning its parents about the atrocities of the Second World War. Had Auschwitz been an anomaly, they wanted to know, or is there a Nazi hiding in each of us?

In *Lord of the Flies*, William Golding intimated the latter and scored an instant hit. So much so, argued the influential critic Lionel Trilling, that the novel 'Marked a mutation in culture.'[4] Eventually, Golding even won a Nobel Prize for his oeuvre. His work 'illuminate[s] the human condition in the world of today,' wrote the Swedish Nobel committee, 'with the perspicuity of realistic narrative art and the diversity and universality of myth.'

These days, *Lord of the Flies* is read as far more than 'just' a novel. Sure, it's a made-up story shelved with all the other fiction, but Golding's take on human nature has also made it the veritable textbook on veneer theory. Before Golding, nobody had ever attempted such raw realism in a book about children. Instead of sentimental tales of houses on prairies or lonely little princes, here – ostensibly – was a harsh look at what kids are *really* like.

2

I first read *Lord of the Flies* as a teenager. I remember feeling disillusioned afterwards, as I turned it over and over in my mind. But not for a second did I think to doubt Golding's view of human nature.

That didn't happen until I picked up the book again years later. When I began delving into the author's life, I learned what an unhappy individual he'd been. An alcoholic. Prone to depression. A man who, as a teacher, once divided his pupils into gangs and encouraged them to attack one another. 'I have always understood the Nazis,' Golding confessed, 'because I am of that sort by nature.' And it was 'partly out of that sad self-knowledge' that he wrote *Lord of the Flies*.[5]

Other people held little interest for Golding. As his biographer observes, he didn't even take the trouble to spell acquaintances' names correctly. '[A] more urgent matter to me than actually meeting people,' Golding said, was 'the nature of Man with a capital M.'[6]

And so I began to wonder: had anyone ever studied what real children would do if they found themselves alone on a deserted island? I wrote an article on the subject, in which I compared *Lord of the Flies* to modern scientific insights and concluded that, in all probability, kids would act very differently.[7] I cited biologist Frans de Waal, who said, 'there is no shred of evidence that this is what children left to their own devices will do'.[8]

Readers of that piece responded sceptically. All my examples concerned kids at home, at school, or at summer camp. They didn't answer the fundamental question: what happens when kids are left on a deserted island *all alone*?

Thus began my quest for a real-life *Lord of the Flies*.

Of course, the chances that any university would ever have permitted researchers to leave juvenile test subjects alone in the wilderness for months on end were slim, even in the 1950s. But couldn't it have happened somewhere, sometime, by accident? Say, after a shipwreck?

I started with a basic internet search: 'Kids shipwrecked.' 'Real-life *Lord of the Flies*.' 'Children on an island.' The first hits I got were about a horrid British reality show from 2008 that pitted participants against each other. But after trawling the web for a while, I came across an obscure blog that told an arresting story: 'One day, in 1977, six boys set out from Tonga on a fishing trip. [...] Caught in a huge storm, the boys were shipwrecked on a deserted island. What do they do, this little tribe? They made a pact never to quarrel.'[9]

The article did not provide any sources. After a couple more hours of clicking, I discovered that the story came from a book by a well-known anarchist, Colin Ward, entitled *The Child in The Country* (1988). Ward, in turn, cited a report by an Italian politician, Susanna Agnelli, compiled for some international committee or other.

Feeling hopeful, I went in search of the report. And luck was on my side: I turned up a copy at a second-hand bookshop in the UK. Two weeks later it landed on my doormat. Flicking through, I found what I was looking for on page 94.

Six boys alone on an island. Same story, same details, same wording, and – once again – no source.[10]

Okay, I thought, maybe I can track down this Susanna Agnelli and ask where she got the story. But no such luck: she'd passed away in 2009. If this had really happened, I reasoned, there must be an article about it from 1977. Not only that, the boys could still be alive. But search as I might in archive upon archive, I couldn't find a thing.

Sometimes all it takes is a stroke of luck. Sifting through a newspaper archive one day, I typed a year incorrectly and ended up deep in the 1960s. And there it was. The reference to 1977 in Agnelli's report turns out to have been a typo.

In the 6 October 1966 edition of Australian newspaper *The Age*, a headline jumped out at me: 'Sunday Showing for Tongan Castaways'. The story concerned six boys who had been found three weeks earlier on a rocky islet south of Tonga, an island group in the Pacific Ocean (and a British protectorate until 1970). The boys had been rescued by an Australian sea captain after being marooned on the island of 'Ata for more than a year.

According to the article, the captain had even got a television station to film a re-enactment of the boys' adventure.

'Their survival story already is regarded as one of the great classic stories of the sea,' the piece concluded.

I was bursting with questions. Were the boys still alive? And could I find the television footage? Most importantly, though, I had a lead: the captain's name was Peter Warner. Maybe he was even still alive! But how do you go about locating an elderly man on the other side of the globe?

As I did a search for the captain's name, I had another stroke of luck. In a recent issue of the *Daily Mercury*, a tiny local paper from Mackay, Australia, I came across the headline: MATES SHARE 50-YEAR BOND. Printed alongside was a small photograph of two men, smiling, one with his arm slung around the other. The article began: 'Deep in a banana plantation at Tullera, near Lismore, sits an unlikely pair of mates [...] These men have laughing eyes and a sparkling energy that belies their age. The elder is 83 years old, the son of a wealthy industrialist. The younger, 67, was, literally, a child of nature.'[11]

Their names? Peter Warner and Mano Totau. And where had they met?

On a deserted island.

3

We set out one September morning. My wife Maartje and I had rented a car in Brisbane, on Australia's east coast, and I was seated anxiously at the wheel. The nerves may have had something to do with the fact that it took me six attempts to pass my driving test, and now I had to navigate on the left-hand side of the road. But also: I was on my way to meet a main character from 'one of the great classic stories of the sea'.

Some three hours later we arrived at our destination, a spot in the middle of nowhere that stumped Google Maps. Yet there he was, sitting out in front of a low-slung house off this dirt road: the man who rescued six lost boys fifty years ago. Captain Peter Warner.

Before I tell his story, there are a few things you should know about Peter, because his life alone is worth a movie. He was born the youngest son of Arthur Warner, once one of the richest and most powerful men in Australia. Back in the 1930s, Arthur ruled over a vast empire called Electronic Industries, which dominated the country's radio market at the time.

Peter had been groomed to follow in his father's footsteps. Instead, at the age of seventeen, he ran away. He went to sea in search of adventure. 'I'd prefer to fight nature rather than human beings,' he later explained.[12]

Peter spent the next few years sailing the seven seas, from Hong Kong to Stockholm, from Shanghai to St Petersburg. When he finally returned five years later, the prodigal son proudly presented his father with a Swedish captain's certificate. Unimpressed, Warner Sr demanded his son learn a useful profession.

'What's easiest?' Peter asked.

'Accountancy,' Arthur lied.[13]

It took another five years of night school for Peter to earn his degree. He went to work for his father's company, yet the sea still beckoned, and whenever he could get away Peter went to Tasmania, where he kept his own fishing fleet. It was this fishing on the side that brought him to Tonga in the winter of 1966. He had arranged an audience with the king to ask permission to trap lobster in Tongan waters. Unfortunately, His Majesty Taufa'ahau Tupou IV refused.

Disappointed, Peter headed back to Tasmania, but on the way he took a little detour, outside royal waters, to cast his nets. And that's when he saw it: a minuscule island in the azure sea.

The island of 'Ata.

Peter knew that no ships had anchored there in ages. The island had been inhabited once, up until one dark day in 1863, when a slave ship appeared on the horizon and sailed off with the natives. Since then, 'Ata had been deserted – cursed and forgotten.

But Peter noticed something odd. Peering through his binoculars, he saw burned patches on the green cliffs. 'In the tropics it's unusual for fires to start spontaneously,' he told us, a half century later. 'So I decided to investigate.' As his boat approached the western tip of the island, Peter heard a shout from the crow's nest.

'Someone's calling!' yelled one of his men.

'Nonsense,' Peter shouted back. 'It's just squawking seabirds.'

But then, through his binoculars, he saw a boy. Naked. Hair down to his shoulders. This wild creature leaped from the cliffside and plunged into the water. Suddenly more boys followed, screaming at the top of their lungs.

Peter ordered his crew to load their guns, mindful of the Polynesian custom of dumping dangerous criminals on remote islands. It didn't take long for the first boy to reach the boat. 'My name is Stephen,' he cried in perfect English. 'There are six of us and we reckon we've been here fifteen months.'

Peter was more than a little sceptical. The boys, once aboard, claimed they were students at a boarding school in Nuku'alofa, the Tongan capital. Sick of school meals, they had decided to take a fishing boat out one day, only to get caught in a storm.

Likely story, Peter thought. Using his two-way radio, he called in to Nuku'alofa. 'I've got six kids here,' he told the operator. 'If I give you their names, can you telephone the school to find out if they're pupils there?'

'Stand by,' came the response.

Twenty minutes ticked by. (As Peter tells this part of the story, he gets a little misty-eyed.)

Finally, 'A very tearful operator came on the radio, and said, "You found them! These boys have been given up for dead. Funerals have been held. If it's them, this is a miracle!"'

I asked Peter if he'd ever heard of the book *Lord of the Flies*.

'Yes, I've read it,' he laughed. 'But that's a completely different story!'

4

In the months that followed I tried to reconstruct as precisely as possible what had happened on that tiny island of 'Ata. Peter's memory turned out to be excellent. Even at the age of ninety, everything he recounted was consistent with the other sources.[14]

My foremost other source lived a few hours' drive from Peter. Mano Totau, fifteen years old at the time and now pushing seventy, counted the captain among his closest friends. A couple days after our visit with Peter, Mano was waiting to welcome me and my wife in his garage in Deception Bay, just north of Brisbane.

The real *Lord of the Flies*, Mano told us, began in June 1965.

The protagonists were six boys, all pupils at St Andrew's, a strict Anglican boarding school in Nuku'alofa. The oldest was sixteen, the youngest thirteen, and they had one main thing in common: they were bored witless. The teenagers longed for adventure instead of assignments, for life at sea instead of school.

So they came up with a plan to escape: to Fiji, some five hundred miles away, or even all the way to New Zealand. 'Lots of other kids at school knew about it,' Mano recalled, 'but they all thought it was a joke.'

There was only one obstacle. None of them owned a boat, so they decided to 'borrow' one from Mr Taniela Uhila, a fisherman they all disliked.

The boys took little time to prepare for the voyage. Two sacks of bananas, a few coconuts and a small gas burner were all the supplies they packed. It didn't occur to any of them to bring a map, let alone a compass. And none of them was an experienced sailor. Only the youngest, David, knew how to steer a boat (which, according to him, 'was why they wanted me to come along').[15]

The journey began without a hitch. No one noticed the small craft leaving the harbour that evening. Skies were fair; only a mild breeze ruffled the calm sea.

But that night the boys made a grave error. They fell asleep. A few hours later they awoke to water crashing down over their heads. It was dark. All they could see were foaming waves cresting around them. They hoisted the sail, which the wind promptly tore to shreds. Next to break was the rudder. 'When we get home,' joked Sione, the eldest, 'we must tell Taniela his boat is just like himself – too old and cranky.'[16]

In the days that followed there was little to joke about. 'We drifted for eight days,' Mano told me. 'Without food. Without water.' The boys tried catching fish. They managed to collect some rainwater in hollowed-out coconut shells and shared it equally between them, each taking a sip in the morning and another in the evening. Sione tried boiling seawater on the gas burner, but it tipped over and burned a large part of his leg.

Then, on the eighth day, they spied a miracle on the horizon. Land. A small island, to be precise. Not a tropical paradise with waving palm trees and sandy beaches, but a hulking mass of rock, jutting up more than a thousand feet out of the ocean.

Eight days adrift in the Pacific

Path of the six boys' journey to 'Ata

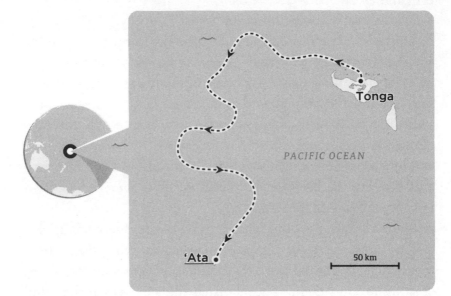

These days, 'Ata is considered uninhabitable. A rugged Spanish adventurer found this out a few years ago. He thought it might be a good spot for the shipwreck expeditions he organises for rich folk with unusual needs. He went to check it out, but just nine days in the poor guy had to call it quits. When a journalist asked if his company would be expanding to the rocky outcrop, he answered in no uncertain terms.

'Never. The island is far too tough.'[17]

The teenagers had a rather different experience. 'By the time we arrived,' Captain Warner wrote in his memoirs, 'the boys had set up a small commune with food garden, hollowed-out tree trunks to store rainwater, a gymnasium with curious weights, a badminton court, chicken pens and a permanent fire, all from handiwork, an old knife blade and much determination.'[18]

It was Stephen – later an engineer – who, after countless failed attempts, managed to produce a spark using two sticks. While the boys in the make-believe *Lord of the Flies* come to blows over the fire, those in the real-life *Lord of the Flies* tended their flame so it never went out, for more than a year.

The kids agreed to work in teams of two, drawing up a strict roster for garden, kitchen and guard duty. Sometimes they quarrelled, but whenever that happened they solved it by imposing a time-out. The squabblers would go to opposite ends of the island to cool their tempers, and, 'After four hours or so,' Mano later remembered, 'we'd bring them back together. Then we'd say 'Okay, now apologise.' That's how we stayed friends.'[19]

Their days began and ended with song and prayer. Kolo fashioned a makeshift guitar from a piece of driftwood, half a coconut shell and six steel wires salvaged from their wrecked boat – an instrument Peter has kept all these years – and played it to help lift their spirits.

And their spirits needed lifting. All summer long it hardly rained, driving the boys frantic with thirst. They tried constructing a raft in order to leave the island, but it fell apart in the crashing surf.[20] Then there was the storm that swept across the island and dropped a tree on their hut.

Worst of all, Stephen slipped one day, fell off a cliff and broke his leg. The other boys picked their way down after him and then helped him back up to the top. They set his leg using sticks and leaves. 'Don't worry,' Sione joked. 'We'll do your work, while you lie there like King Taufa'ahau Tupou himself!'[21]

The boys were finally rescued on Sunday 11 September 1966.

Physically, they were in peak condition. The local physician, Dr Posesi Fonua, later expressed astonishment at their muscled physiques and Stephen's perfectly healed leg.

But this wasn't the end of the boys' little adventure, because, when they arrived back in Nuku'alofa, they found the police waiting to meet them. You might expect the officers to have been thrilled at the return of the town's six lost sons. But no. They boarded Peter's boat, arrested the boys and threw them in jail. Mr Taniela Uhila, whose sailing boat the boys had 'borrowed' fifteen months earlier, was still furious, and he'd decided to press charges.

Fortunately for the boys, Peter came up with a plan. It occurred to him that the story of their shipwreck was perfect Hollywood material. Six kids marooned on an island ... it was a tale people would be talking about for years. And being his father's corporate accountant, Peter managed the company's movie rights and knew people in television.[22]

The captain knew exactly what to do. First, from Tonga, he called up the manager of Channel 7 in Sydney. 'You can have the Australian rights,' he told them. 'Give me the world rights. Then we'll spring these kids out of prison and take them back to the island.' Next, Peter went around to see Mr Uhila and paid him £150 for his old boat, getting the boys released on the condition that they would cooperate with the movie.

A few days later, a team from Channel 7 arrived in the ancient DC-3 that flew a once-weekly service to Tonga. Describing the scene to Maartje and me, Peter chuckles. 'Out of that aircraft stepped three of those TV types in their city suits and pointy shoes.'

By the time the group arrived on 'Ata with the six boys in tow, the gang from Channel 7 was green around the gills. Worse, they didn't know how to swim. 'Don't worry,' Peter assured them. 'These boys will save you.'

The captain rowed the trembling men out to the breakers. 'This is where you get out.'

Even fifty years later, the memory brings tears to Peter's eyes – this time from laughter. 'So I tossed them out, and these Australian television people were sinking down, the Tongans diving down, picking them up, taking them through the surf, bashing them up against the rocks.'

Next, the group had to scale the cliff, which took the rest of the day. When they finally arrived at the top, the TV crew collapsed, exhausted. Not surprisingly, the documentary about 'Ata was no success. Not only were the shots lousy, but most of the 16mm film went missing, leaving a grand total of only thirty minutes. 'Actually,' Peter amends, 'twenty minutes, plus commercials.'

Naturally, as soon as I heard about the Channel 7 documentary I wanted to see it. Peter didn't have it, so back in the Netherlands I contacted an agency that specialised in tracking down and restoring old recordings. But search as they might, it was nowhere to be found.

Then Peter intervened again, putting me in touch with an independent filmmaker named Steve Bowman who had visited the 'boys' in 2006. Steve was frustrated that their story had never got the attention it deserved. His own documentary had never aired because his distributor went bankrupt, but he still had his raw interviews. He kindly offered to share them with me and also to put me in touch with Sione, the oldest of the bunch. Then he announced that he had the sole remaining copy of that original 16mm documentary.

'May I see it?' I asked Steve.

'Of course,' he answered.

And that's how – months after stumbling across a story on an obscure blog about six shipwrecked kids – I was suddenly watching the original 1966 footage on my laptop. 'I am Sione Fataua,' it began. 'Five classmates from St Andrew's High School and I washed ashore on this island in June 1965.'

The mood when the boys returned to their families in Tonga was jubilant. Almost the entire island of Ha'afeva – population nine hundred – had turned out to welcome them home. 'No sooner did one party end than preparations for another began,' narrated the 1966 documentary voiceover.

Peter was proclaimed a national hero. Soon he received a message from King Taufa'ahau Tupou IV himself, inviting the captain for another audience. 'Thank you for rescuing six of my subjects,' His Royal Highness said. 'Now, is there anything I can do for you?'

The captain didn't have to think long. 'Yes! I would like to trap lobster in these waters and start a business here.'

This time the king consented. Peter returned to Sydney, resigned from his father's company and commissioned a new ship. Then he had the six boys brought over and granted them the thing that had started it all: an opportunity to see the world beyond Tonga. He hired Sione, Stephen, Kolo, David, Luke and Mano as the crew of his new fishing boat.

The name of the boat? The *Ata*.

5

This is the real-life *Lord of the Flies*.

Turns out, it's a heart-warming story – the stuff of bestselling novels, Broadway plays and blockbuster movies.

It's also a story that nobody knows. While the boys of 'Ata have been consigned to obscurity, William Golding's book is still widely read. Media historians even credit him as being the unwitting originator of one of the most popular entertainment genres on television today: reality TV.

The premise of so-called reality shows, from *Big Brother* to *Temptation Island*, is that human beings, when left to their own devices, behave like beasts. 'I read and re-read *Lord of the Flies*,' divulged the creator of hit series *Survivor* in an interview. 'I read it first when I was about twelve, again when I was about twenty and again when I was thirty and since we did the programme as well.'[23]

The show that launched this whole genre was MTV's *The Real World*. Since it first aired in 1992, every episode opens with a cast member reciting, 'This is the true story of seven strangers [...] Find out what happens when people stop being polite and start getting real.'

Lying, cheating, provoking, antagonising – these are what each instalment would have us believe it means to 'get real'. But take the time to look behind the scenes of programmes like these and you'll see candidates being led on, boozed up and played off against each other in ways that are nothing less than shocking. It shows just how much manipulation it takes to bring out the worst in people.

Another reality show, *Kid Nation*, once tried throwing forty kids together in a ghost town in New Mexico in the hope they would wind up at each other's throats. That didn't happen. 'Periodically they would find that we were getting along too well,' one participant later recalled, 'and they'd have to induce something for us to fight over.'[24]

You could say: What does it really matter? We all know it's just entertainment.

But seldom is a story only a story. Stories can also be nocebos. In a recent study, psychologist Bryan Gibson demonstrated that watching *Lord of the Flies*-type television can make people more aggressive.[25] In children, the correlation between seeing violent images and aggression in adulthood is stronger than the correlation between asbestos and cancer, or between calcium intake and bone mass.[26]

Cynical stories have an even more marked effect on the way we look at the world. In Britain, another study demonstrated that girls who watch more reality TV also more often say that being mean and telling lies are necessary to get ahead in life.[27] As media scientist George Gerbner summed up: '[He] who tells the stories of a culture really governs human behaviour.'[28]

It's time we told a different kind of story.

The real *Lord of the Flies* is a story of friendship and loyalty, a story that illustrates how much stronger we are if we can lean on each other. Of course, it's only one story. But if we're going to make *Lord of the Flies* required reading for millions of teenagers, then let's also tell them about the time real kids found themselves stranded on a deserted island. 'I used their survival story in our social studies classes,' one of the boys' teachers at St Andrew's High School in Tonga recalled years later. 'My students couldn't get enough of it.'[29]

So what happened to Peter and Mano? If you happen to find yourself on a banana plantation outside Tullera, near Lismore, you may well run into them: two older men, trading jokes, arms draped around each other's shoulders. One the son of a big industrialist, the other from more humble roots. Friends for life.

After my wife took Peter's picture, he turned to a cabinet and rummaged around for a bit, then drew out a heavy stack of papers that he laid in my hands. His memoirs, he explained, written for his children and grandchildren.

I looked down at the first page. 'Life has taught me a great deal,' it began, 'including the lesson that you should always look for what is good and positive in people.'

Peter Warner, September 2017. Photo © Maartje ter Horst.

Mano Totau, September 2017. Photo © Maartje ter Horst.

Part One

THE STATE
OF NATURE

'Mankind are so much the same, in all times and places, that history informs us of nothing new or strange in this particular. Its chief use is only to discover the constant and universal principles of human nature.'

David Hume (1711–1776)

Is that heart-warming story of six boys on the island of 'Ata an aberration? Or does it signify something more profound? Is it an isolated anecdote, or an exemplary illustration of human nature?

Are we humans, in other words, more inclined to be good or evil?

It's a question philosophers have grappled with for hundreds of years. Consider the Englishman Thomas Hobbes (1588–1679), whose *Leviathan* set off a shockwave when it was published in 1651. Hobbes was censured, condemned and castigated, and yet we still know his name, while his criticasters are long forgotten. My edition of *The Oxford History of Western Philosophy* describes his magnum opus as 'the greatest work of political philosophy ever written'.

Or take the French philosopher Jean-Jacques Rousseau (1712–78), who penned a succession of volumes that got him into ever-deeper trouble. He was condemned, his books were burned and a warrant was issued for his arrest. But while the names of all his petty persecutors are lost to memory, Rousseau remains known to this day.

The two never met. By the time Rousseau was born, Hobbes had been dead thirty-three years. Nevertheless, they continue to be pitted against each other in the philosophical boxing ring. In one corner is Hobbes: the pessimist who would have us believe in the wickedness of human nature. The man who asserted that civil society alone could save us from our baser instincts. In the other corner, Rousseau: the man who declared that in our heart of hearts we're all good. Far from being our salvation, Rousseau believed 'civilisation' is what ruins us.

Even if you've never heard of them, the opposing views of these two heavyweights are at the root of society's deepest divides. I know of no other debate with stakes as high, or ramifications as far-reaching. Harsher punishments versus better social services, art school versus reform school, top-down management versus empowered teams, old-fashioned breadwinners versus baby-toting dads – take just about any debate you can think of and it goes back, in some way, to the opposition between Hobbes and Rousseau.

Let's begin with Thomas Hobbes. He was one of the first philosophers to argue that if we really want to know ourselves, we have to understand how our ancestors lived. Imagine we were to travel back 50,000 years in time. How did we interact in those hunting and gathering days? How did we conduct ourselves when there was no code of law, no courts or judges, no prisons or police?

Hobbes thought he knew. 'Read thyself,' he wrote: dissect your own fears and emotions and you will 'thereby read and know what are the thoughts and passions of all other men upon the like occasions.'

When Hobbes applied this method to himself, the diagnosis he made was bleak indeed.

Back in the old days, he wrote, we were free. We could do whatever we pleased, and the consequences were horrific. Human life in that state of nature was, in his words, 'solitary, poor, nasty, brutish, and short'. The reason, he theorised, was simple. Human beings are driven by fear. Fear of the other. Fear of death. We long for safety and have 'a perpetual and restless desire of power after power, that ceaseth only in death'.

The result? According to Hobbes, 'a condition of war of all against all.' *Bellum omnium in omnes.*

But don't worry, he assured us. Anarchy can be tamed and peace established – if we all just agree to relinquish our liberty. To put ourselves, body and soul, into the hands of a solitary sovereign. He named this absolute ruler after a biblical sea monster: the Leviathan.

Hobbes' thinking provided the basic philosophical rationale for an argument that would be repeated thousands, nay, millions of times after him, by directors and dictators, governors and generals …

'Give us power, or all is lost.'

Fast-forward about a hundred years and we encounter Jean-Jacques Rousseau one day, a no-name musician, walking to the prison at Vincennes, just outside Paris. He's on his way to visit his friend Denis Diderot, a poor philosopher who's been locked up for cracking an unfortunate joke about the mistress of a government minister.

And that's when it happens. Having paused to rest underneath a shady tree, Rousseau is leafing through the latest issue of *Mercure de France* when his eye falls on an advertisement that will change his life. It's a call for submissions to an essay contest being held by the Academy of Dijon. Entrants are instructed to answer the following question:

'Has the restoration of the sciences and arts contributed to the purification of morals?'

Rousseau immediately knows his answer. 'At the moment of that reading,' he later wrote, 'I beheld another universe and became another man.' In that instant, he realised that civil society is not a blessing, but a curse. As he continued on his way to where his innocent friend was incarcerated, he understood 'that man is naturally good, and that it is from these institutions alone that men become wicked'.

Rousseau's essay won first prize.

In the years that followed, he grew to become one of the leading philosophers of his day. And, I have to say, his work is still a delight to read. Not only was Rousseau a great thinker, he was a gifted writer, too. Take this scathing passage about the invention of private property:

The first man, who, after enclosing a piece of ground, took it into his head to say, 'This is mine,' and found people simple enough to believe him, was the true founder of civil society. How many crimes, how many wars, how many murders, how many misfortunes and horrors, would that man have saved the human species, who pulling up the stakes or filling up the ditches should have cried to his fellows: Be sure not to listen to this imposter; you are lost, if you forget that the fruits of the earth belong equally to us all, and the earth itself to nobody!

Ever since the birth of that cursed civil society, Rousseau argued, things had gone wrong. Farming, urbanisation, statehood – they hadn't lifted us out of chaos, but enslaved and doomed us. The invention of writing and the printing press had only made matters worse. 'Thanks to typographic characters,' he wrote, 'the dangerous reveries of Hobbes [...] will remain for ever.'

In the good old days before bureaucrats and kings, Rousseau believed that everything was better. Back when humans existed in a 'state of nature' we were still compassionate beings. Now we'd become cynical and self-interested. Once we'd been healthy and strong. Now we were indolent and feeble. Civilisation had, to his mind, been one giant mistake. We should never have squandered our freedom.

Rousseau's thinking provided the basic philosophical rationale for an argument that would be repeated thousands, nay, millions of times, after him, by anarchists and agitators, free spirits and firebrands:

'Give us liberty, or all is lost.'

So here we are, three hundred years later.

Few other philosophers have had as profound an impact on our politics, education and world view as these two. The whole science of economics became premised on the Hobbesian notion of human nature, which sees us as rational, self-serving individuals. Rousseau, for his part, has been enormously influential in education, due to his belief – revolutionary in the eighteenth century – that children should grow up free and unfettered.

To this day, the influence of Hobbes and Rousseau is staggering. Our modern camps of conservative and progressive, of realists and idealists, can be traced back to them. Whenever an idealist advocates more freedom and equality, Rousseau beams down approvingly. Whenever the cynic grumbles that this will only spark more violence, Hobbes nods in agreement.

The writings of these two do not make for light reading. Rousseau, in particular, leaves lots of room for interpretation. But these days we're in a position to test their principal point of contention. Hobbes and Rousseau, after all, were armchair theorists, while we've been gathering scientific evidence for decades now.

In Part 1 of this book I'll examine the question: which philosopher was right? Should we be grateful that our days of nature are behind us? Or were we once noble savages?

A great deal hinges on the answer.

3

The Rise of *Homo puppy*

I

The first thing to understand about the human race is that, in evolutionary terms, we're babies. As a species we've only just emerged. Imagine that the whole history of life on earth spans just one calendar year, instead of four billion. Up until about mid-October, bacteria had the place to themselves. Not until November did life as we know it appear, with buds and branches, bones and brains.

And we humans? We made our entrance on 31 December, at approximately 11 p.m. Then we spent about an hour roaming around as hunter-gatherers, only getting around to inventing farming at 11:58 p.m. Everything else we call 'history' happened in the final sixty seconds to midnight: all the pyramids and castles, the knights and ladies, the steam engines and rocket ships.

In the blink of an eye, *Homo sapiens* populated the entire globe, from its coldest tundras to its hottest deserts. We even became the first species to blast off the planet and set foot on the moon.

But why us? Why wasn't the first astronaut a banana? Or a cow? Or a chimpanzee?

These may sound like silly questions. But genetically we're 60 per cent identical to bananas, 80 per cent indistinguishable from cows and 99 per cent the same as chimpanzees. It's not exactly a given that we would milk cows, instead of them milking us, or that we would cage chimps and not the other way around. Why should that 1 per cent make all the difference?

History of life on Earth (4,000 million years)
Represented as one calendar year

First life on earth

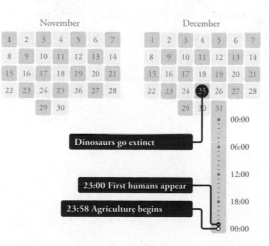

January

1	2	3	4	5	6	7
8	9	10	11	12	13	14
15	16	17	18	19	20	21
22	23	24	25	26	27	28
		29	30	31		

February

1	2	3	4	5	6	7
8	9	10	11	12	13	14
15	16	17	18	19	20	21
22	23	24	25	26	27	28

March

1	2	3	4	5	6	7
8	9	10	11	12	13	14
15	16	17	18	19	20	21
22	23	24	25	26	27	28
		29	30	31		

April

1	2	3	4	5	6	7
8	9	10	11	12	13	14
15	16	17	18	19	20	21
22	23	24	25	26	27	28
		29	30			

May

1	2	3	4	5	6	7
8	9	10	11	12	13	14
15	16	17	18	19	20	21
22	23	24	25	26	27	28
		29	30	31		

June

1	2	3	4	5	6	7
8	9	10	11	12	13	14
15	16	17	18	19	20	21
22	23	24	25	26	27	28
		29	30			

July

1	2	3	4	5	6	7
8	9	10	11	12	13	14
15	16	17	18	19	20	21
22	23	24	25	26	27	28
		29	30	31		

August

1	2	3	4	5	6	7
8	9	10	11	12	13	14
15	16	17	18	19	20	21
22	23	24	25	26	27	28
		29	30	31		

September

1	2	3	4	5	6	7
8	9	10	11	12	13	14
15	16	17	18	19	20	21
22	23	24	25	26	27	28
		29	30			

October

1	2	3	4	5	6	7
8	9	10	11	12	13	14
15	16	17	18	19	20	21
22	23	24	25	26	27	28
		29	30	31		

November

1	2	3	4	5	6	7
8	9	10	11	12	13	14
15	16	17	18	19	20	21
22	23	24	25	26	27	28
		29	30			

December

1	2	3	4	5	6	7
8	9	10	11	12	13	14
15	16	17	18	19	20	21
22	23	24	25	26	27	28
		29	30	31		

00:00
06:00
12:00
18:00
00:00

Dinosaurs go extinct

23:00 First humans appear

23:58 Agriculture begins

For a long time we considered our privileged position to be part of God's plan. The human race was better, smarter and superior to every other living thing – the pinnacle of His creation.

But imagine, again, that ten million years ago (on roughly 30 December), aliens visited the earth. Could they have predicted the rise of *Homo sapiens*? Not a chance. The genus *Homo* didn't yet exist. The earth was literally still a planet of the apes, and certainly nobody was building cities, writing books, or launching rockets.

The uncomfortable truth is that we, too – the creatures that consider ourselves so unique – are the product of a blind process called evolution. We belong to a raucous family of mostly hairy creatures also known as primates. Right up to ten minutes before midnight, we even had other hominins for company.[1] Until they mysteriously disappeared.

I distinctly remember when I first began to grasp the significance of evolution. I was nineteen and listening to a lecture about Charles Darwin on my iPod. I was depressed for a week. Sure, I'd learned about the English scientist as a kid, but I'd attended a Christian school and the biology teacher presented evolution as just another wacky theory. Um, not exactly, I would later learn.

The basic ingredients for the evolution of life are straightforward. You need:

Lots of suffering.
Lots of struggle.
Lots of time.

In short, the process of evolution comes down to this: animals have more offspring than they can feed. Those that are slightly better adapted to their environment (think thicker fur or better

camouflage) have a slightly higher chance of surviving to pro-
create. Now imagine a friendly game of run till you're dead,
in which billions upon billions of creatures bite the dust, some
before they can pass the baton to their offspring. Keep this
footrace going long enough – say four billion years – and the
minuscule variations between parents and children can branch
out into a vast and varied tree of life.

That's it. Simple, but brilliant.

For Darwin the biologist, who'd once considered becoming
a priest, the impossibility of reconciling the cruelty of nature
with the biblical story of creation ultimately destroyed his faith
in God. Consider, he wrote, the parasitoid wasp, an insect that
lays its eggs in a live caterpillar. Upon hatching, the larvae
eat the caterpillar from the inside out, inducing a horrific,
drawn-out death.

What kind of sick mind would think up something like that?

Nobody, that's who. There is no mastermind, no grand
design. Pain, suffering and struggle are merely the engines of
evolution. Can you blame Darwin for putting off publishing
his theory for years? Writing to a friend, he said it was 'like
confessing a murder'.[2]

Evolutionary theory doesn't seem to have got any jollier
since. In 1976, British biologist Richard Dawkins published his
magnum opus on the instrumental role genes play in the evolution of life, tellingly titled *The Selfish Gene*. It's a depressing
read. Are you counting on nature to make the world a better
place? Then Dawkins is clear: Don't hold your breath. 'Let us
try to teach generosity and altruism,' he writes, 'because we are
born selfish.'[3]

Forty years after its publication, the British public voted *The
Selfish Gene* the most influential science book ever written.[4]
But countless readers felt dispirited upon reaching the end. 'It
presents an appallingly pessimistic view of human nature [...]

yet I cannot present any arguments to refute its point of view,' wrote one. 'I wish I could unread it.'[5]

So here we are, *Homo sapiens*, the product of a brutish and protracted process. While 99.9 per cent of species have gone extinct, we're still here. We've conquered the planet and – who knows? – the Milky Way could be next.

But why us?

You might assume that it's because our genes are the most selfish of them all. Because we are strong and smart, lean and mean. And yet ... are we? As for being strong: no, not really. A chimpanzee can clobber us without breaking sweat. A bull can effortlessly lance us with one sharp horn. At birth we're utterly helpless, and after that we remain frail, slow, and not even all that good at escaping up trees.

Maybe it's because we're so clever? On the face of it, you might think so. *Homo sapiens* has a whopper of a brain that guzzles energy like a sauna at the North Pole. Our brains may account for just 2 per cent of our body weight, but they use 20 per cent of the calories we consume.[6]

But are human beings really all that brilliant? When we do a difficult sum or draw a pretty picture, we've usually learned that skill from someone else. Personally, for example, I can count to ten. Impressive, sure, but I doubt I could have come up with a numeric system by myself.

Scientists have been trying for years to figure out which animal has the most natural smarts. The standard procedure is to compare our intelligence to that of other primates like orangutans and chimpanzees. (Normally, the human subjects are toddlers, since they've had less time to crib off other people.) A good example is the series of thirty-eight tests designed by a research team in Germany, which assesses subjects on spatial awareness, calculation and causality.[7] The chart below shows the results.

How smart are humans really?

Scores on three intelligence tests

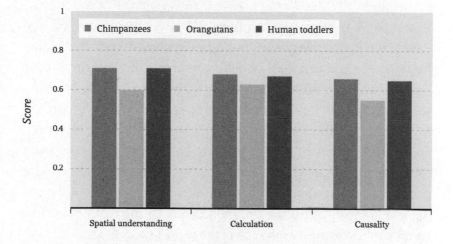

That's right, toddlers score the same as animals at the zoo. And it gets worse. Turns out that our working memory and speed of processing information – traditionally regarded as one of the cornerstones of human intelligence – won't be winning us any prizes either.

This was demonstrated by Japanese researchers who developed a test to assess how adults stack up against chimpanzees. Subjects were placed in front of a screen that flashed a set of digits (from one to nine). After a given amount of time – always less than a second – the digits were replaced by white squares. Test subjects were instructed to tap the spots on the screen where the numbers had appeared, in order from low to high.

Briefly, it looked like Team Human would beat Team Chimp. But when researchers made the test *harder* (by having the numbers disappear sooner), the chimps pulled ahead. The Einstein of the group was Ayuma, who was faster than the other participants and made fewer errors.[8] Ayuma was a chimpanzee.

Okay, judged on raw brain power, humans do no better than our hairier cousins. So, then, what are we using our great big brains for?

Maybe we're more cunning. That's the crux of the 'Machiavellian intelligence' hypothesis, named after the Italian Renaissance philosopher Niccolò Machiavelli, author of *The Prince* (1513). In this handbook for rulers, Machiavelli counsels weaving a web of lies and deception to stay in power. According to adherents of this hypothesis, that's precisely what we've been doing for millions of years: devising ever more inventive ways to swindle one another. And because telling lies takes more cognitive energy than being truthful, our brains grew like the nuclear arsenals of Russia and the US during the Cold War. The result of this mental arms race is the *sapien* superbrain.

If this hypothesis were true, you'd expect humans to beat other primates handily in games that hinge on conning your opponent. But no such luck. Numerous studies show that chimps outscore us on these tests and that humans are lousy liars.[9] Not only that, we're predisposed to trust others, which explains how con artists can fool their marks.[10]

This brings me to another odd quirk of *Homo sapiens*. Machiavelli, in his classic book, advises never revealing your emotions. Work on your poker face, he urges; shame serves no purpose. The object is to win, by fair means or foul. But if only the shameless win, why are humans one of the only species in the whole animal kingdom to *blush*?

Blushing, said Charles Darwin, is 'the most peculiar and the most human of all expressions'. Wanting to know if this phenomenon was universal, he sent letters to everyone in his foreign network, polling missionaries, merchants, and colonial bureaucrats.[11] Yes, they all replied, people here blush, too.

But why? Why didn't blushing die out?

2

It's August 1856. At a limestone quarry north of Cologne, two workers have just made the discovery of a lifetime. They've uncovered the skeleton of one of the most controversial creatures ever to walk the earth.

Not that they realise it. Old bones, mostly bear or hyena, routinely crop up in their line of work and just get thrown out with the other waste. But this time their overseer notices the remains lying in the dump. Thinking they may be the bones of a cave bear, he decides they would make a cool gift for Johann Carl Fuhlrott, a science teacher at the local high school. Like many people in the days before Netflix, Fuhlrott is an avid fossil collector.

As soon as he lays eyes on them, Fuhlrott realises these are no ordinary bones. At first he thinks the skeleton is human, but something isn't right. The skull is strange. It's sloping and elongated, with a jutting brow ridge and a nose that's too big.

That week, the local papers report on the astounding discovery of a 'Race of Flatheads' in the Neander Valley. A professor at the University of Bonn, Hermann Schaaffhausen, reads about the find and contacts Fuhlrott. They arrange to meet – the amateur and the pro – and exchange notes. A few hours later the two are in agreement:

The bones belong to not just any human, but to a whole different *species* of human.

'These bones are *antediluvian*,' Fuhlrott declares.[12] That is, they predate the Great Flood, which makes them the remains of a creature that lived before God inundated the earth.

It's hard to overstate just how shocking this conclusion was at the time. Pure heresy. When Fuhlrott and Schaaffhausen announce their findings at a gathering of the erudite Lower Rhine Society for Science and Medicine, they meet with stunned disbelief.[13] Ridiculous, shouts an anatomy professor, this is the skeleton of a Russian Cossack who died in the Napoleonic Wars. Nonsense, calls another, it's just 'some poor fool or recluse' whose head is misshapen from disease.[14]

But then more bones turn up. All over Europe, museums dive into their collections and resurface with more of the oblong skulls. At first, they get dismissed as malformations, then it begins to dawn on scientists that this could indeed be a whole different kind of human. Before long, someone dubs the species: *Homo stupidus*.[15] His 'thoughts and desires,' expounds a respected anatomist, 'never soared beyond those of a brute.'[16] The classification recorded in the annals of science is more subtle, and refers to the valley where the bones were found.

Homo neanderthalensis.

To this day, the popular image of the Neanderthal is of a stupid lout, and it's not hard to fathom why. We have to face the uncomfortable fact that, until not long ago, our species shared the planet with other kinds of humans.

Skulls of *Homo sapiens* and *Homo neanderthalensis* compared

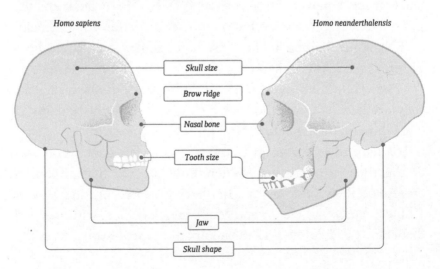

Homo sapiens   Homo neanderthalensis

Skull size

Brow ridge

Nasal bone

Tooth size

Jaw

Skull shape

Scientists have determined that 50,000 years ago there were at least five hominins besides us – *Homo erectus, Homo floresiensis, Homo luzonensis, Homo denisova* and *Homo neanderthalensis* – all of them humans, just as the goldfinch, the house finch and the bullfinch are all finches. So besides the question of why we put chimps in the zoo instead of the other way around, there's another mystery: what happened to the 'Race of Flatheads'? What did we do with our other *Homo* brothers and sisters? Why did they all disappear?

Was it that the Neanderthals were weaker than us? On the contrary, they were the proto-muscleman, with biceps like Popeye after downing a can of spinach. More importantly, they were tough. That's what two American archaeologists ascertained in the 1990s after detailed analysis of a vast number of Neanderthal bone fractures. This led them to draw parallels with a modern occupational group that also suffers a high rate of 'violent encounters' with large animals. Rodeo cowboys.

The archaeologists got in touch with the – I kid you not – Professional Rodeo Cowboys Association, which in the 1980s had registered 2,593 injuries among its members.[17] Comparing this data with that from Neanderthals, they found striking similarities. The only difference? Neanderthals weren't riding bucking broncos and roping cattle, but spearing mammoths and sabre-toothed cats.[18]

Okay, so if they weren't weaker, maybe Neanderthals were dumber than us?

Here things get more painful. The Neanderthal brain was, on average, 15 per cent larger than our brains now: 1,500 cm^3 versus 1,300 cm^3. We may boast a superbrain, they packed a gigabrain. We have a Macbook Air, and they got the Macbook Pro.

As scientists continue to make new discoveries about Neanderthals, the growing consensus is that this species was astoundingly intelligent.[19] They built fires and cooked food. They made clothing, musical instruments, jewellery and cave paintings. There are even indications that we borrowed some inventions from the Neanderthals, like certain stone tools, and possibly even the practice of burying the dead.

So what gives? How did Neanderthals, with their brawn, big brains, and ability to survive two whole ice ages, end up getting wiped off the earth? Having managed to stick it out for more than 200,000 years, why was it game over for the Neanderthals soon after *Homo sapiens* arrived on the scene?

There is one final and much more sinister hypothesis.

If we weren't stronger, or more courageous, or smarter than the Neanderthals, maybe we were just meaner. 'It may well be,' speculates Israeli historian Yuval Noah Harari, 'that when *Sapiens* encountered Neanderthals, the result was the first and most significant ethnic-cleansing campaign in history'.[20] Pulitzer Prize-winning geographer Jared Diamond concurs: 'murderers have been convicted on weaker circumstantial evidence'.[21]

3

Could it be true? Did we wipe out our hominin cousins?

Flash forward to the spring of 1958. Lyudmila Trut, a biology student at Moscow State University, comes knocking at the office door of Professor Dmitri Belyaev. He's a zoologist and geneticist and is looking for someone to run an ambitious new research programme. She's still in school, but determined to land the job.[22]

The professor is kind and courteous. At a time when the Soviet scientific establishment mostly takes a condescending attitude towards women, Dmitri treats Lyudmila as an equal. And he decides to let her in on his secret plan. This plan will require her to travel to Siberia, to a remote location near the border with Kazakhstan and Mongolia, where the professor is launching an experiment.

He cautions Lyudmila to think carefully before she agrees, because this venture is dangerous. The communist regime has stamped evolutionary theory as a lie propagated by capitalists and has banned genetic research of any kind. Ten years earlier, they executed Dmitri's older brother, also a geneticist. For this reason, the team will present the experiment to the outside world as a study on precious fox pelts.

In reality, it's about something altogether different. 'He told me,' Lyudmila said years later, 'that he wanted to make a dog out of a fox.'[23]

What the young scientist didn't realise was that she had just agreed to embark on an epic quest. Together, Dmitri Belyaev and Lyudmila Trut would unravel the very origins of humankind.

They started out with a very different question: how do you turn a fierce predator into a friendly pet? A hundred years earlier, Charles Darwin had already noted that domesticated animals – pigs, rabbits, sheep – present some remarkable similarities. For starters, they're a few sizes smaller than their wild forebears. They have smaller brains and teeth and often floppy ears, curly tails, or white-spotted fur. Perhaps most interesting of all, they retain some juvenile traits their whole lives.

This was a puzzle that had perplexed Dmitri for years. Why do domesticated animals look the way they do? Why did all those innumerable farmers, all those innumerable years ago, prefer puppies and piglets with corkscrew tails, droopy ears and baby faces, and breed them for these particular traits?

The Russian geneticist had a radical hypothesis. He suspected these cute features were merely *by-products* of something else, a metamorphosis that happens organically if over a sufficiently long period of time animals are consistently selected for one specific quality:

Friendliness.

So this was Dmitri's plan. He wanted to replicate within a couple of decades what had taken nature millennia to produce. He wanted to turn wild animals into pet material, simply by breeding only the most amiable individuals. For his test case, Dmitri chose the silver fox, an animal never domesticated and so viciously aggressive that the researchers could only handle them wearing elbow-length gloves two inches thick.

Dmitri warned Lyudmila not to get her hopes up. The experiment would take years, maybe even a lifetime, most likely with nothing to show for the effort. But Lyudmila didn't need to think twice. A few weeks later, she boarded the Trans-Siberian Express.

The fox breeding farm Dmitri contracted turned out to be a vast complex, the thousands of cages emitting a cacophony of howling. Even with everything she'd read on the behaviour of silver foxes, Lyudmila was not prepared for how ferocious they were in person. She started making her rounds past all the cages that first week. Wearing protective gloves, she would reach a hand inside to see how the animals reacted. If she sensed the slightest hesitation, Lyudmila selected that fox for breeding.

In retrospect, it's remarkable how quickly it all happened.

In 1964, with the experiment in its fourth generation, Lyudmila saw the first fox wag its tail. To ensure that any such behaviours were indeed the result of natural selection (and were not acquired), Lyudmila and her team had kept all contact with the animals to a minimum. But that became increasingly difficult: within a few generations, the foxes were literally begging for attention. And who could say no to a drooling, tail-wagging fox cub?

In the wild, foxes become significantly more aggressive at about eight weeks old, but Lyudmila's selectively bred foxes remained permanently juvenile, preferring nothing more than to play all day long. 'These tamer foxes,' Lyudmila later wrote, 'seemed to be resisting the mandate to grow up.'[24]

Meanwhile, there were noticeable physical changes, too. The foxes' ears dropped. Their tails curled and spots appeared on their coats. Their snouts got shorter, their bones thinner and the males increasingly resembled the females. The foxes even began to bark, like dogs. And before long they were responding when the keepers called their names – behaviour never before seen in foxes.

And remember, none of these were traits for which Lyudmila had selected. Her only criterion had been friendliness – all the other characteristics were just by-products.

Dmitri Belyaev with his silver foxes, Novosibirsk, 1984. Dmitri died the following year, but his research program continues to this day. Source: Alamy.

By 1978, twenty years after this experiment began, a lot had changed in Russia. No longer did biologists have to conceal their research. The theory of evolution was not a capitalist plot after all, and the Politburo was now keen to promote Russian science.

In August that year, Dmitri managed to arrange for the International Congress of Genetics to be hosted in Moscow. Guests were received at the State Kremlin Palace – capacity 6,000 – where the champagne flowed freely and there was plenty of caviar to go around.

But none of that impressed the guests nearly as much as Dmitri's talk. After a brief introduction, the lights dimmed and a video began to play. Onto the screen bounded an unlikely creature: a silver fox, tail wagging. A chorus of exclamations arose from the audience, and the excited chatter continued long after the lights came back up.

But Dmitri wasn't finished yet. In the hour that followed, he set out his revolutionary idea. He suspected, he said, that the changes in these foxes had everything to do with hormones. The more amiable foxes produced fewer stress hormones and more serotonin (the 'happy hormone') and oxytocin (the 'love hormone').

And one last thing, Dimitri said in closing. This didn't apply only to foxes.

The theory 'can also, of course, apply to human beings'.[25]

Looking back, it was a historic statement.

Two years after Richard Dawkins published his bestseller about egoistic genes, concluding that people are 'born selfish,' here was an unknown Russian geneticist claiming the opposite. Dmitri Belyaev's theory was that people are domesticated apes. That for tens of thousands of years, the nicest humans had the most kids. That the evolution of our species, in short, was predicated on 'survival of the friendliest'.

If Dmitri was right, our own bodies should hold clues to

prove this theory. Like pigs, rabbits, and now silver foxes, human beings should have got smaller and cuter.

Dmitri had no way to test his hypothesis, but science has since advanced. When in 2014 an American team began looking at human skulls from a range of periods over the past 200,000 years, they were able to trace a pattern.[26] Our faces and bodies have grown considerably softer, more youthful and more feminine, they found. Our brains have shrunk by at least 10 per cent, and our teeth and jawbones have become, to use the anatomical jargon, *paedomorphic*. In plain English: childlike.

If you compare our heads to those of Neanderthals, the differences are even more pronounced. We have shorter and rounder skulls, with a smaller brow ridge. What dogs are to wolves, we are to Neanderthals.[27] And just as mature dogs look like wolf puppies, humans evolved to look like baby monkeys.

Meet *Homo puppy*.

**The domestication
of humans and dogs**

Resulted in:
· Friendlier behavior
· More serotonin and oxytocin
· Longer juvenile stage
· More feminine and juvenile appearance
· Better communication

Source: Brian Hare, "Survival of the Friendliest," *Annual Review of Psychology* (2017).

This transformation in our appearance accelerated roughly fifty thousand years ago. Intriguingly, that's around the same time Neanderthals disappeared and we came up with a slew of new inventions – like better sharpening stones, fishing lines, bows and arrows, canoes and cave paintings. None of this appears to make evolutionary sense. People got weaker, more vulnerable, and more infantile. Our brains got smaller, yet our world got more complex.

How come? And how was *Homo puppy* able to conquer the world?

4

Who better to answer this question than a true puppy expert? Growing up in Atlanta in the 1980s, Brian Hare was crazy about dogs. He decided to study biology, only to find out that biologists weren't too interested in dogs. After all, canines may be cute, but they're not all that clever.

In college, Brian took a class with Michael Tomasello, a professor in developmental psychology who would become his mentor and colleague. Tomasello's research focused on chimps, a species generally deemed far more interesting than dogs. During his sophomore year, Brian, then just nineteen, assisted in administering an intelligence test.

It was a classic *object choice test* in which a tasty treat is hidden and subjects are given hints about where to find it. Human toddlers ace this test, but it stumps chimpanzees. No matter how emphatically Professor Tomasello and his students pointed to the spot where they had hidden a banana, the apes remained clueless.

After another long day of gesticulating, Brian blurted out, 'I think my dog can do it.'

'Sure,' his professor smirked.

'No, really,' Brian insisted. 'I bet he could pass the tests.'[28]

Twenty years later, Brian Hare is himself a professor in evolutionary anthropology. Using a series of meticulous experiments, he's been able to demonstrate that dogs are incredibly intelligent, in some instances even smarter than chimpanzees (despite dogs' smaller brains).

At first scientists didn't understand this one bit. How could *dogs* be intelligent enough to pass the object choice test? They certainly hadn't inherited their brains from their wolf ancestors, because wolves score just as poorly on Brian's test as orangutans and chimpanzees. And they didn't pick it up from their owners, because puppies can pass the test at nine weeks old.

Brian's colleague and adviser, the primatologist Richard Wrangham, suggested that canine intelligence might arise *on its own*, as a chance by-product, like corkscrew tails and drop ears. But Brian didn't buy that; how could a trait as instrumental as social intelligence be an accident? Rather, the young biologist suspected that our ancestors had selectively bred the smartest dogs.

There was only one way Brian could test his suspicion. It was time for a trip to Siberia. Years earlier, Brian had read about an obscure study by a Russian geneticist who purportedly had turned foxes into dogs. By the time Brian stepped off the Trans-Siberian Express, in 2003, Lyudmila and her team had already bred forty-five generations. Brian would be the first foreign scientist to study the silver foxes, and he started with the object choice test.

If his hypothesis was correct, the friendly foxes and the ferocious foxes would flunk the test in equal measure, since Dmitri and Lyudmila had bred them on the basis of friendliness, *not* intelligence. If Brian's adviser Richard was right, and intelligence was a coincidental by-product of friendliness, then the selectively bred foxes would pass the test with flying colours.

Long story short: the results supported the by-product theory and proved Brian wrong. The latest generation of friendly foxes was not only remarkably astute, but also much smarter than their aggressive counterparts. As Brian put it, 'The foxes totally rocked my world.'[29]

Up until then the assumption had always been that domestication diminishes brainpower, literally reducing grey matter and in the process sacrificing skills needed to survive in the wild. We all know the clichés. Sly as a fox. Dumb as an ox. But Brian came to a completely different conclusion. 'If you want a clever fox,' he says, 'you don't select for cleverness. You select for friendliness.'[30]

5

This brings us back to the question I posed at the beginning of this chapter. What makes human beings unique? Why do we build museums, while the Neanderthals are stuck in the displays?

Let's take another look at the results of those thirty-eight tests done with primates and toddlers. What I neglected to mention earlier is that subjects were also assessed on a fourth skill: social learning. That is, the ability to learn from others. And the results of this last test reveal something interesting.

Humans' true superpower

Scores on four intelligence tests

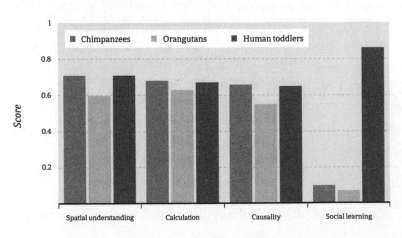

This figure perfectly illustrates the skill that sets humans apart. Chimpanzees and orangutans score on a par with human two-year-olds on almost every cognitive test. But when it comes to learning, the toddlers win hands down. Most kids score 100 per cent, most apes 0.

Human beings, it turns out, are ultrasocial learning machines. We're born to learn, to bond and to play. Maybe it's not so strange, then, that blushing is the only human expression that's uniquely human. Blushing, after all, is quintessentially social – it's people showing they care what others think, which fosters trust and enables cooperation.

Something similar happens when we look one another in the eye, because humans have another weird feature: we have whites in our eyes. This unique trait lets us follow the direction of other people's gazes. Every other primate, more than two hundred species in all, produces melanin that tints their eyes. Like poker players wearing shades, this obscures the direction of their gaze.

But not humans. We're open books; the object of our attention plain for all to see. Imagine how different human friendships and romance would be if we couldn't look each other in the eye. How would we feel able to trust one another? Brian Hare suspects our unusual eyes are another product of human domestication. As we evolved to become more social, we also began revealing more about our inner thoughts and emotions.[31]

Add to this the smoothing of our large brow ridge, the *torus supraorbitalis* seen in Neanderthal skulls and in living chimpanzees and orangutans. Scientists think the protruding ridge may have impeded communication, because we now use our eyebrows in all kinds of subtle ways.[32] Just try expressing surprise, sympathy or disgust and notice how much your eyebrows do.

Humans, in short, are anything but poker-faced. We constantly leak emotions and are hardwired to relate to the people around us. But far from being a handicap, this is our true superpower, because sociable people aren't only more fun to be around, in the end they're smarter, too.

The best way to conceptualise this is to imagine a planet inhabited by two tribes: the Geniuses and the Copycats. The Geniuses are brilliant, and one in ten of them invents something truly amazing (say, a fishing rod) at some point in their lives. The Copycats are less cognitively endowed, so only one out of every thousand eventually teaches him or herself to fish. That makes the Geniuses a hundred times smarter than the Copycats.

But the Geniuses have a problem. They're not all that social. On average, the Genius who invents a fishing rod has only one friend they can teach to fish. The Copycats have on average ten friends, making them ten times as social.

Now let's assume that teaching someone else to fish is tricky and only succeeds half the time. The question is: which group profits from the invention most? The answer, calculates anthropologist Joseph Henrich, is that only one in five Geniuses will ever learn to fish, half having figured it out on their own and the other half learning it from somebody else. By contrast, although a mere 0.1 per cent of the Copycats will work out the technique on their own, 99.9 per cent of them will end up able to fish, because they'll pick it up from other Copycats.[33]

Neanderthals were a little like the Geniuses. Their brains were bigger individually, but collectively they weren't as bright. On his own, a *Homo neanderthalensis* may have been smarter than any one *Homo sapiens*, but the *sapiens* cohabited in larger groups, migrated from one group to another more frequently and may also have been better imitators. If Neanderthals were a super-fast computer, we were an old-fashioned PC – *with* wi-fi. We were slower, but better connected.

Some scientists theorise that the development of human language, too, is a product of our sociability.[34] Language is an excellent example of a system that Copycats might not think up themselves but can learn from one another, over time giving rise to talking humans in much the same way that Lyudmila's foxes began to bark.

So what happened to the Neanderthals? Did *Homo puppy* wipe them out after all?

This notion may make for a thrilling read or documentary, but there's not a shred of archaeology to support it. The more plausible theory is that we humans were better able to cope with the harsh climatic conditions of the last ice age (115,000–15,000 years ago) because we'd developed the ability to work together.

And that depressing book *The Selfish Gene*? It fit right in with 1970s-era thinking – a time hailed as the 'Me Decade' by the *New York* magazine. In the late 1990s, an avid Richard Dawkins fan decided to put his take on Dawkins ideas into practice. Rather than making him feel pessimistic, the book inspired CEO Jeffrey Skilling to run an entire corporation – the energy giant Enron – on the mechanism of greed.

Skilling set up a 'Rank & Yank' system for performance reviews at Enron. A score of 1 placed you among the company's top performers and gave you a fat bonus. A score of 5 put you at the bottom, a group 'sent to Siberia' – besides being humiliated, if you couldn't find another position within two weeks you were fired. The result was a Hobbesian business culture with cut-throat competition between employees. In late 2001 the news broke that Enron had been engaging in massive accounting fraud. When the dust finally settled, Skilling was in prison.

These days, 60 per cent of the largest US corporations still employ some variation of the Rank & Yank system.[35] 'It is a Hobbesian universe,' journalist Joris Luyendijk said of London's financial services sector in the aftermath of the 2008 credit crisis, 'of all against all, with relationships that are characteristically nasty, brutish and short.'[36] The same goes for organisations like Amazon and Uber, which systematically pit their workers against each other. Uber, in the words of one anonymous employee, is a 'Hobbesian jungle' where 'you can never get ahead unless someone else dies'.[37]

Science has advanced considerably since the 1970s. In subsequent editions of *The Selfish Gene*, Richard Dawkins scrapped his assertions about humans' innate selfishness, and the theory has lost credence with biologists. Although struggle and competition clearly factor into the evolution of life, every first-year biology student now learns that cooperation is much more critical.

This is a truth as old as the hills. Our distant ancestors knew the importance of the collective and rarely idolised individuals. Hunter-gatherers the world over, from the coldest tundras to the hottest deserts, believed that everything is connected. They saw themselves as a part of something much bigger, linked to all other animals, plants and Mother Earth. Perhaps they understood the human condition better than we do today.[38]

Is it any wonder, then, that loneliness can quite literally make us sick? That a lack of human contact is comparable to smoking fifteen cigarettes a day?[39] That having a pet lowers our risk of depression?[40] Human beings crave togetherness and interaction.[41] Our spirits yearn for connection just as our bodies hunger for food. It's that longing, more than anything else, that enabled *Homo puppy* to shoot for the moon.

When I understood this, the notion of evolution didn't feel like such a downer any more. Maybe there's no creator and no cosmic plan. Maybe our existence is just a fluke, after millions of years of blind fumbling. But at least we're not alone. We have each other.

4

Colonel Marshall and the Soldiers Who Wouldn't Shoot

I

And now for the elephant in the room.

We humans also have a dark side. Sometimes, *Homo puppy* does horrific things unprecedented in the animal kingdom. Canaries don't run prison camps. Crocodiles don't build gas chambers. Never in all of history has a koala felt impelled to count, lock up and wipe out a whole race of fellow creatures. These crimes are singularly human. So besides being exceptionally prosocial, *Homo puppy* can also be shockingly cruel. Why?

It seems we have to face a painful fact. 'The mechanism that makes us the kindest species,' says Brian Hare, puppy expert, 'also makes us the cruelest species on the planet.'[1] People are social animals, but we have a fatal flaw: we feel more affinity for those who are most like us.

This instinct seems to be encoded in our DNA. Take the hormone oxytocin, which biologists have long known plays a key role in childbirth and breastfeeding. When they first discovered that the hormone is also instrumental in romance, there was a flurry of excitement. Spray a little oxytocin up your noses, some conjectured, and have the best date ever.

In fact, why not have crop dusters mist the masses? Oxytocin — which Lyudmila Trut's cute Siberian foxes showed high levels of — makes us kinder, gentler, more laid-back and serene. It transforms even the biggest jerk into a friendly puppy. That's

74

why it's often touted in mushy terms such as the 'milk of human kindness' and the 'hug hormone'.

But then came another newsflash. In 2010, researchers at the University of Amsterdam found that the effects of oxytocin seem limited to one's own group.[2] The hormone not only enhances affection for friends, it can also intensify aversion to strangers. Turns out oxytocin doesn't fuel universal fraternity. It powers feelings of 'my people first'.

2

Maybe Thomas Hobbes was right after all.

Maybe our prehistory *was* 'a war of all against all'. Not among friends, but between enemies. Not with those we knew, but with strangers we didn't. If that's true, then by now archaeologists should have found innumerable artefacts of our aggression, and their excavations surely would have uncovered evidence that we're hardwired for war.

I'm afraid they have. The first such clues were unearthed in 1924, when a miner dislodged the skull of a small, apelike individual in north-western South Africa, outside the village of Taung. This skull wound up in the hands of anatomist Raymond Dart. He identified it as *Australopithecus africanus*, one of the first hominins to walk the earth – two, possibly three million years ago.

From the beginning, Dart was disturbed by his discovery. Studying this skull and the bones of our other ancestors, he saw numerous injuries. What had caused them? His conclusion wasn't pretty. These early hominins must have used stones, tusks and horns to kill their prey, said Dart, and by the looks of the remains, animals weren't their only victims. They also murdered each other.

Raymond Dart became one of the first scientists to characterise human beings as innately bloodthirsty cannibals, and his
'killer ape theory' made headlines around the world. It was only
with the advent of farming a mere 10,000 years ago, he said, that
we switched to a more compassionate diet. The very incipience
of our civilisation could be the reason for our 'widespread reluctance' to acknowledge what, deep down, we truly are.[3]

Dart himself had no such qualms. Our earliest ancestors were,
he wrote, 'confirmed killers: carnivorous creatures, that seized
living quarries by violence, battered them to death, tore apart
their broken bodies, dismembered them limb from limb, slaking
their ravenous thirst with the hot blood of victims and greedily
devouring livid writhing flesh'.[4]

Now that Dart had laid the groundwork, it was open season
for science, and ranks of researchers followed in his footsteps.
First up was Jane Goodall, who studied our chimp cousins in
Tanzania. Given that chimps were long regarded as peaceable
plant eaters, it came as a massive shock to Goodall when in 1974
she arrived in the middle of an all-out ape war.

For four years, two groups of chimpanzees fought a brutal
battle. Appalled, Goodall long kept her discovery under wraps,
and when she did finally share her findings with the world,
many people didn't believe her. She described scenes of chimps
'cupping the victim's head as he lay bleeding with blood pouring
from his nose and drinking the blood, twisting a limb, tearing
pieces of skin with their teeth'.[5]

One of Goodall's students, a primatologist by the name of
Richard Wrangham (and advisor to puppy expert Brian Hare
from Chapter 3), speculated in the 1990s that our ancestors must
have been a kind of chimpanzee. Tracing a direct line from those
predatory primates to the battlegrounds of the twentieth century, Wrangham surmised that war was simply in our blood,

making 'modern humans the dazed survivors of a continuous, five-million-year habit of lethal aggression'.[6]

What led him to this verdict? Simple: the killers survive, the softies die. Chimpanzees have a penchant for ganging up and ambushing solitary peers, much as bullies take out their baser instincts on school playgrounds.

You may be thinking: that's all well and good, but these scientists were talking about chimps and other apes. Isn't *Homo puppy* unique? Didn't we conquer the world precisely because we're so affable? What does the record actually show about the days when *we* were still hunting and gathering?

Early studies seemed to put us in the clear. In 1959, the anthropologist Elizabeth Marshall Thomas published a book about the !Kung people, who still live in the Kalahari Desert today.[7] Title? *The Harmless People*. Its message dovetailed with the spirit of the 1960s, when a new generation of left-leaning scientists came on the anthropological scene, keen to give our ancestors a Rousseauian makeover. Anyone wanting to know how we lived in the past, they asserted, need only look at nomads still foraging in the present.

Thomas and her colleagues showed that although there was the occasional rumble in the jungle or on the savanna, these tribal 'wars' amounted to little more than name-calling. Sometimes, someone let fly an arrow, but if one or two warriors got injured the tribes usually called it a day. *See?* said the progressive academics, Rousseau was right; cavemen really were noble savages.

Sadly for the hippies, however, counter-evidence swiftly began piling up.

More focused research by later anthropologists determined that the killer-ape theory held true for hunter-gatherers, too. Their ritual battles may look innocent enough, but the bloody attacks under cover of night and the massacres of men, women and children aren't so easily explained away. Even the !Kung, on closer inspection, proved to be fairly bloodthirsty, if you observed them long enough. (And the murder rate plummeted after !Kung territory came under state control in the 1960s. That is, when Hobbes's *Leviathan* arrived to impose the rule of law.)[8]

And this was just the beginning. In 1968, the anthropologist Napoleon Chagnon came along with a study on the Yanomami people of Venezuela and Brazil that really shook things up. Title? *The Fierce People*. It described a society 'in a chronic state of war'. Worse still, it revealed that men who were killers also had more wives and children – makes sense then that violence is in our blood.

But the argument wasn't truly settled until 2011, with the publication of Steven Pinker's monumental book *The Better Angels of Our Nature*. It's the magnum opus of a psychologist who was already ranked among the world's most influential intellectuals: a massive doorstop of a book with 802 pages in extra-small font and packed with graphs and tables. Perfect for knocking your enemies out cold.

'Today,' writes Pinker, 'we can switch from narratives to numbers.'[9] And those numbers speak for themselves. Average share of skeletons from twenty-one archaeological sites that show signs of a violent death? Fifteen per cent. Average share of deaths caused by violence in eight tribes still foraging today? Fourteen per cent. Average over the whole twentieth century, including two world wars? Three per cent. Same average now?

One per cent.

'We started off nasty,' Pinker concurs with Hobbes.[10] Biology, anthropology and archaeology all point in the same direction: humans may be nice to their friends, we're cold-blooded when it comes to outsiders. In fact, we're the most warmongering creatures on the planet. Fortunately, Pinker reassures his readers, we've been ennobled by the 'artifices of civilization'.[11] The invention of farming, writing and the state have served to rein in our aggressive instincts, applying a thick coat of civilisation over our nasty, brutish nature.

Under the weight of all the statistics trotted out in this hefty tome, the case seemed closed. For years, I thought Steven Pinker was right, and Rousseau cracked. After all, the results were in and numbers don't lie.

Then I found out about Colonel Marshall.

3

It's 22 November 1943. Night has fallen on an island in the Pacific, and the Battle of Makin has just begun. The offensive is unfolding as planned when something strange happens.[12]

Samuel Marshall, colonel and historian, is there to see it. He's accompanied the first American contingent ashore as they try to take the island, which is in Japanese hands. Rarely has a historian been this close to the action. The invasion itself is a perfectly isolated operation, almost like a lab experiment. It's the ideal opportunity for Marshall to observe how war plays out in real time.

The men advance three miles that day in the blistering heat. When they halt in the evening, nobody has the energy to dig themselves in, so they don't realise there's a Japanese camp a short distance away. The attack starts after dark. Japanese forces storm the American position, making eleven attempts in all. Despite being outnumbered they nearly manage to break through American lines.

The next day, Marshall wonders what went wrong. He knows there's only so much you can learn by peering at flags on a map or reading the officers' logbooks. So he decides to do something that's never been tried. It's revolutionary in the world of historical scholarship. That same morning, he rounds up the American soldiers and interviews them in groups. He asks them to speak freely, allowing lower ranks to disagree with their superiors.

As far as strategies go, it's genius. 'Marshall almost immediately realized he had stumbled onto the secret of accurate combat reporting,' a colleague would later write. 'Every man remembered something – a piece to be fitted into the jigsaw puzzle.'[13] And that's how the colonel makes a baffling discovery.

Most of the soldiers never fired their guns.

For centuries, even millennia, generals and governors, artists and poets had taken it for granted that soldiers fight. That if there's one thing that brings out the hunter in us, it's war. War is when we humans get to do what we're so good at. War is when we shoot to kill.

But as Colonel Samuel Marshall continued to interview groups of servicemen, in the Pacific and later in the European theatre, he found that only 15 to 25 per cent of them had actually fired their weapons. At the critical moment, the vast majority balked. One frustrated officer related how he had gone up and down the lines yelling, 'Goddammit! Start shooting!' Yet, 'they fired only while I watched them or while some other officer stood over them'.[14]

The situation on Makin that night had been do-or-die, when you would expect everyone to fight for their lives. But in his battalion of more than three hundred soldiers, Marshall could identify only thirty-six who actually pulled the trigger.

Was it a lack of experience? Nope. There didn't seem to be any difference between new recruits and experienced pros when it came to willingness to shoot. And many of the men who didn't fire had been crack shots in training.

Maybe they just chickened out? Hardly. Soldiers who didn't fire stayed at their posts, which meant they ran as much of a risk. To a man, they were courageous, loyal patriots, prepared to sacrifice their lives for their comrades. And yet, when it came down to it, they shirked their duty.

They failed to shoot.

In the years after the Second World War, Samuel Marshall would become one of the most respected historians of his generation. When he spoke, the US Army listened. In his 1946 book *Men Against Fire* – still read at military academies to this day – he stressed that 'the average and normally healthy individual [...] has such an inner usually unrealized resistance toward killing a fellow man that he will not of his own volition take life'.[15] Most people, he wrote, have a 'fear of aggression' that is a normal part of our 'emotional make-up'.[16]

What was going on? Had the colonel uncovered some powerful instinct? Published when veneer theory was at its peak and Raymond Dart's killer ape model all the rage, Marshall's findings were hard to take in. The colonel had a hunch that his analysis wasn't limited to Allied servicemen in the Second World War, but applied to *all soldiers throughout history*. From the Greeks at Troy to the Germans at Verdun.

Though Marshall enjoyed a distinguished reputation during his lifetime, in the 1980s the doubts began to surface. 'Pivotal S. L. A. Marshall Book on Warfare Assailed as False,' declared the front page of the *New York Times* on 19 February 1989. The magazine *American Heritage* went so far as to call it a hoax, alleging that Marshall had 'made the whole thing up' and never conducted any group interviews at all. 'That guy perverted history' a former officer scoffed. 'He didn't understand human nature.'[17]

Marshall was unable to defend himself, having died twelve years earlier. Other historians then dived into the fray – and into the archives – and found indications that Marshall had indeed twisted the facts at times. But the group interviews had been real enough, and he certainly asked soldiers if they'd fired their M1s.[18]

After days of reading Marshall, his detractors and his defenders, I no longer knew what to think. Was I just a little too eager for the colonel to be right? Or was he really onto something? The deeper I delved into the controversy, the more Marshall struck me as an intuitive thinker – not a stellar statistician, granted, but definitely a perceptive observer.

The big question was: is there any further evidence to back him up?

Short answer? Yes.

Long answer? Over the last decades, proof that Colonel Marshall was right has been piling up.

First of all, colleagues on the front observed the same thing as Marshall. Lieutenant Colonel Lionel Wigram complained during the 1943 campaign in Sicily that he could rely on no more than a quarter of his troops.[19] Or take General Bernard Montgomery, who in a letter home wrote, 'The trouble with our British boys is that they are not killers by nature.'[20]

When historians later began interviewing veterans of the Second World War, they found that more than half had never killed anybody, and most casualties were the work of a small minority of soldiers.[21] In the US Air Force, less than 1 per cent of fighter pilots were responsible for almost 40 per cent of the planes brought down.[22] Most pilots, one historian noted, 'never shot anyone down or even tried to'.[23]

Prompted by these findings, scholars began revisiting assumptions about other wars as well. Such as the 1863 Battle of Gettysburg at the height of the American Civil War. Inspection of the 27,574 muskets recovered afterwards from the battlefield revealed that a staggering 90 per cent were still loaded.[24] This made no sense at all. On average, a rifleman spent 95 per cent of the time loading his gun and 5 per cent firing it. Since priming a musket for use required a whole series of steps (tear open the cartridge with your teeth, pour gunpowder down the barrel, insert the ball, ram it in, put the percussion cap in place, cock back the hammer and pull the trigger), it was strange, to say the least, that so many guns were still *fully* loaded.

But it gets even stranger. Some twelve thousand muskets were double-loaded, and half of those more than triple. One rifle even had *twenty-three* balls in the barrel – which is absurd. These soldiers had been thoroughly drilled by their officers. Muskets, they all knew, were designed to discharge one ball at a time.

So what were they doing? Only much later did historians figure it out: loading a gun is the perfect excuse not to shoot it. And if it happened to be loaded already, well, you just loaded it again. And again.[25]

Similar findings were made in the French army. In a detailed survey conducted among his officers in the 1860s, French colonel Ardant du Picq discovered that soldiers are not all that into fighting. When they did fire their weapons, they often aimed too high. That could go on for hours: two armies emptying their rifles over each other's heads, while everyone scrambled for an excuse to do something else – anything else – in the meantime (replenish ammo, load your weapon, seek cover, whatever).

'The obvious conclusion,' writes military expert Dave Grossman, 'is that most soldiers were *not* trying to kill the enemy.'[26]

Reading this, I suddenly recalled a passage about the very same phenomenon by one of my favourite authors. 'In this war everyone always did miss everyone else, when it was humanly possible,' wrote George Orwell in his Spanish Civil War classic, *Homage to Catalonia*.[27] This is not to imply there were no casualties, of course; but according to Orwell, most soldiers who wound up in the infirmary had injured themselves. By accident.

In recent years, a steady stream of experts has rallied behind Colonel Marshall's conclusions. Among them is sociologist Randall Collins, who analysed hundreds of photographs of soldiers in combat and, echoing Marshall's estimates, calculates that only about 13 to 18 per cent fired their guns.[28]

'The Hobbesian image of humans, judging from the most common evidence, is empirically wrong,' Collins asserts. 'Humans are hardwired for [...] solidarity; and this is what makes violence so difficult.'[29]

4

To this day, our culture is permeated by the myth that it's easy to inflict pain on others. Think about trigger-happy action heroes like Rambo and the ever-fighting Indiana Jones. Look at the way fist fights go on for ever in movies and on TV – where violence spreads like an infection: a character trips, falls on someone else, who lands an accidental punch, and before you know it you're in the middle of a war of all against all.

But the image cooked up by Hollywood has about as much to do with real violence as pornography has with real sex. In reality, says the science, violence isn't contagious, it doesn't last very long and it's anything but easy.

The more I read about Colonel Marshall's analyses and subsequent research, the more I began to doubt the notion of our warmongering nature. After all, if Hobbes was right, we should all take pleasure in killing another person. True, it might not rate as high as sex, but it certainly wouldn't inspire a deep *aversion*.

If, on the other hand, Rousseau was right, then nomadic foragers should have been largely peaceable. In that case, we must have evolved our intrinsic antipathy towards bloodshed over the tens of thousands of years that *Homo puppy* went about populating the earth.

Could Steven Pinker, the psychologist of the weighty tome, be mistaken? Could his seductive statistics about the high human toll of prehistoric wars – that I eagerly cited in earlier books and articles – be wrong?

I decided to go back to square one. This time, I steered clear of publications intended for a popular readership and delved deeper into the academic literature. It wasn't long before I discovered a pattern. When a scientist portrayed humans as homicidal primates, the media was quick to seize on their work. If a colleague argued the reverse, scarcely anyone listened.

This made me wonder: are we being misled by our fascination with horror and spectacle? What if scientific truth is diametrically different to what the bestselling and most-cited publications would have us believe?

Let's revisit Raymond Dart, the man who back in the 1920s examined the first unearthed remains of *Australopithecus africanus*. After inspecting the damaged bones of these two-million-year-old hominins, he decided they must have been bloodthirsty cannibals.

That conclusion was a hit. Just look at movies such as the original *Planet of the Apes* and *2001: A Space Odyssey* (both 1968), which cashed in on killer ape theory. 'I'm interested in the brutal and violent nature of man,' confirmed director Stanley Kubrick in an interview, 'because it's a true picture of him.'[30]

Not until many years later did scientists realise that the forensic remains of *Australopithecus africanus* pointed in an altogether different direction. The bones, experts now agree, were damaged not by other hominins (wielding stones, tusks, or horns), but by predators. So, too, the individual whose skull

Raymond Dart analysed in 1924. In 2006 the new verdict came in: the offender had been a large bird of prey.[31]

What about chimpanzees, our near kin who have been known to tear each other limb from limb? Aren't they living proof that blood lust is baked into our genes?

This continues to be a point of contention. Among other things, scholars disagree on the question *why* chimps go on the attack. Some say human interference itself is to blame, charging that if you regularly feed chimps bananas – like Jane Goodall in Tanzania – it sparks them to be more aggressive. After all, nobody wants to miss out on these treats.[32]

As tantalising as this explanation sounded at first, in the end I wasn't convinced. What clinched it was a big study from 2014 presenting data collected in eighteen chimp colonies over a period of fifty years.[33] No matter which way they looked at it, the researchers could find no correlation between the number of chimpicides and human interference. The chimpanzees, they concluded, were equally capable of savagery without the external stimuli.

Fortunately our family tree has more branches. Gorillas, for example, are far more peaceable. Or, better yet, bonobos. These primates, with their attenuated neck, fine-boned hands and small teeth, prefer to play the day away, are as friendly as can be and never completely grow up.

Ring any bells? Sure enough, biologists suspect that, like *Homo puppy*, the bonobo has domesticated itself. Their faces, incidentally, look uncannily human.[34] If we want to draw parallels, this is where we ought to start.

But how relevant is this heated debate over our nearest kin really? Humans are not chimpanzees, and we're not bonobos either. All told there are more than two hundred different species of primates, with significant variations between them. Robert Sapolsky, a leading primatologist, believes apes have little to teach us about our own human ancestors, saying, 'The debate is an empty one.'[35]

We need to return to the real question – the question that gripped Hobbes and Rousseau.

How violent were the first *human beings?*

Earlier I said there are two ways to find out. One: study modern-day hunter-gatherers living the same life our ancestors did. Two: dig for old bones and other remains our ancestors left behind.

Let's start with number one. I already mentioned Napoleon Chagnon's *The Fierce People*, the bestselling anthropology book of all time. Chagnon showed that the Yanomami people of Venezuela and Brazil have a thing for war, and that homicidal Yanomami men fathered three times as many children as their pacifist counterparts ('wimps', in Chagnon's words[36]).

But how reliable is his research? The current scientific consensus is that most tribes still living the hunter-gatherer life today are not representative of how our ancestors lived. They're up to their ears in civilised society and have frequent contact with agriculturalists and urbanites. The simple fact that they've been followed around by anthropologists renders them 'contaminated' as a study population. (Incidentally, few tribes are more 'contaminated' than the Yanomami. In exchange for their help, Chagnon handed out axes and machetes, then concluded that these people were awfully violent.)[37]

And Chagnon's assertion that killers fathered more children than peaceniks? It literally doesn't add up. That's because he made two serious errors. First, he forgot to correct for age: the killers in his database were on average ten years older than the 'wimps'. So the thirty-five-year-olds had more kids than twenty-five-year-olds. No big surprise there.

87

Chagnon's other fundamental error was that he only included the progeny of killers who were still alive. He disregarded the fact that people who murder other people often get what's coming to them. Revenge, in other words. Ignore these cases and you might as well argue that, after looking solely at the winners, it pays to play the lottery.[38]

After the anthropologist's visit, the Yanomami added a new word to their vocabulary: *anthro*. Definition? 'A powerful non-human with deeply disturbed tendencies and wild eccentricities.'[39] In 1995, this particular *anthro* was barred from ever returning to Yanomami territory.

Clearly, Chagnon's bestseller is best ignored. But we still have psychologist Steven Pinker's eight hundred plus-page testimony with all its graphs and tables as authoritative proof of our violent nature.

In *The Better Angels of Our Nature*, Pinker calculates the average homicide rate among eight primitive societies, arriving at an alarming 14 per cent. This figure appeared in respected journals like *Science* and was endlessly regurgitated by newspapers and on TV. When other scientists took a look at his source material, however, they discovered that Pinker mixed up some things.

This may get a little technical, but we need to understand where he went wrong. The question we want to answer is: which peoples still hunting and gathering today are representative of how humans lived 50,000 years ago? After all, we were nomads for 95 per cent of human history, roving the world in small, relatively egalitarian groups.

Pinker chose to focus almost exclusively on hybrid cultures. These are people who hunt and gather, but who also ride horses or live together in settlements or engage in farming on the side. Now these activities are all relatively recent. Humans didn't start farming until 10,000 years ago and horses weren't domesticated until 5,000 years ago. If you want to figure out how our distant ancestors lived 50,000 years ago, it doesn't make sense to extrapolate from people who keep horses and tend vegetable plots.

But even if we get on board with Pinker's methods, the data is problematic. According to the psychologist, 30 per cent of deaths among the Aché in Paraguay (tribe 1 on his list) and 21 per cent of deaths among the Hiwi in Venezuela and Colombia (tribe 3) are attributable to warfare. These people are out for blood, it would seem.

The anthropologist Douglas Fry was sceptical, however. Reviewing the original sources, he discovered that all forty-six cases of what Pinker categorised as Aché 'war mortality' actually concerned a tribe member listed as 'shot by Paraguayan'.

The Aché were in fact not killing each other, but being 'relentlessly pursued by slave traders and attacked by Paraguayan frontiersmen', reads the original source, whereas they themselves 'desire a peaceful relationship with their more powerful neighbors'. It was the same with the Hiwi. All the men, women and children enumerated by Pinker as war deaths were murdered in 1968 by local cattle ranchers.[40]

There go the iron-clad homicide rates. Far from habitually slaughtering one another, these nomadic foragers were the victims of 'civilised' farmers wielding advanced weaponry. 'Bar charts and numeric tables depicting percentages [...] convey an air of scientific objectivity,' Fry writes. 'But in this case it is all an illusion.'[41]

What can we learn, then, from modern anthropology? What happens if we examine a society that has no settlements, no farming and no horses – a society that can serve as a model for how we once lived?

You guessed it: when we study these types of societies we find that war is a rarity. Based on a list of representative tribes compiled for the journal *Science* in 2013, Douglas Fry concludes that nomadic hunter-gatherers avoid violence.[42] Nomads would rather talk out their conflicts or just move on to the next valley. This sounds a lot like the boys on 'Ata: when tempers flared, they'd head to different parts of the island to cool down.

And another thing. Anthropologists long assumed that pre-historic social networks were small. We wandered through the jungle in bands of thirty or forty relatives, they thought. Any encounters with other groups swiftly devolved into war.

But in 2011 a team of American anthropologists mapped out the social networks of thirty-two primitive societies around the world, from the Nunamuit in Alaska to the Vedda in Sri Lanka. Turns out the nomads are extremely social. They're constantly getting together to eat, party, sing and marry people from other groups.

True, they do their foraging in small teams of thirty to forty individuals, but those groups consist mainly of friends, not family and they're also continually swapping members. As a consequence, foragers have vast social networks. In the case of the Ache in Paraguay and the Hadza in Tanzania, a 2014 study calculated that the average tribe member meets as many as a *thousand* people during his or her lifetime.[43]

In short, there's every reason to think that the average prehistoric human had a large circle of friends. Continually meeting new people meant continually learning new things, and only then could we grow smarter than the Neanderthals.[44]

There is one other way to resolve the question about early man's aggressive nature. By digging. Archaeological evidence may offer the best hope of settling the debate between Hobbes and Rousseau, because the fossil record can't be 'contaminated' by researchers the way tribes can. There's one problem, though: hunter-gatherers travelled light. They didn't have much and they didn't leave much behind.

Fortunately for us, there's an important exception. Cave paintings. If our state of nature was a 'war of all against all' à la Hobbes, then you'd expect that someone, at some point in this period, would have painted a picture of it. But that's never been found. While there are thousands of cave paintings from this time about hunting bison, horses and gazelles, there's not a single depiction of war.[45]

What about ancient skeletons, then? Steven Pinker cites twenty-one excavations having an average murder rate of 15 per cent. But, as before, Pinker's list here is a bit of a mess. Twenty of the twenty-one digs date from a time *after* the invention of farming, the domestication of horses, or the rise of settlements, making them altogether too recent.

So how much archaeological evidence is there for early warfare, before the days of farming, riding horses and living in settled societies? How much proof is there that war is in our nature?

The answer is almost none.

To date, some three thousand *Homo sapiens* skeletons unearthed at four hundred sites are old enough to tell us something about our 'natural state'.[46] Scientists who have studied these sites see no convincing evidence for prehistoric warfare.[47] In later periods, it's a different story. 'War does not go forever backwards in time,' says renowned anthropologist Brian Ferguson. 'It had a beginning.'[48]

5

The Curse of Civilisation

5

The Curse of Civilisation

I

Was Jean-Jacques Rousseau right? Are humans noble by nature, and were we all doing fine until civilisation came along?

I was certainly starting to get that impression. Take the following account recorded in 1492 by a traveller on coming ashore in the Bahamas. He was astonished at how peaceful the inhabitants were. 'They do not bear arms, and do not know them, for I showed them a sword ... and [they] cut themselves out of ignorance.' This gave him an idea. 'They would make fine servants ... With fifty men we could subjugate them all and make them do whatever we want.'

Christopher Columbus – the traveller in question – lost no time putting his plan into action. The following year he returned with seventeen ships and fifteen hundred men, and started the transatlantic slave trade. Half a century later, less than 1 per cent of the original Carib population remained; the rest had succumbed to the horrors of disease and enslavement.

It must have been quite a shock for these so-called savages to encounter such 'civilised' colonists. To some, the very notion that one human being might kidnap or kill another may even have seemed alien. If that sounds like a stretch, consider that there are still places today where murder is inconceivable.

In the vast reaches of the Pacific Ocean, for example, lies a tiny atoll called Ifalik. After the Second World War, the US Navy screened a few Hollywood films on Ifalik to foster goodwill with the Ifalik people. It turned out to be the most appalling thing the islanders had ever seen. The violence on screen so distressed the unsuspecting natives that some fell ill for days.

When years later an anthropologist came to do fieldwork on Ifalik, the natives repeatedly asked her: was it true? Were there *really* people in America who had killed another person?[2]

So at the heart of human history lies this mystery. If we have a deep-seated, instinctive aversion to violence, where did things go wrong? If war had a beginning, what started it?

First, a cautionary note about life in prehistory: we have to guard against painting too romantic a picture of our forebears. Human beings have never been angels. Envy, rage and hatred are age-old emotions that have always taken a toll. In our primal days, resentments could also boil over. And, to be fair, *Homo puppy* would never have conquered the world if we had not, on rare occasions, gone on the offence.

To understand that last point, you need to know something about prehistoric politics. Basically, our ancestors were allergic to inequality. Decisions were group affairs requiring long deliberation in which everybody got to have their say. 'Nomadic foragers,' established one American anthropologist on the basis of a formidable 339 fieldwork studies, 'are universally – and all but obsessively – concerned with being free from the authority of others.'[3]

Power distinctions between people were – if nomads tolerated them at all – temporary and served a purpose. Leaders were more knowledgeable, or skilled, or charismatic. That is, they had the ability to get a given job done. Scientists refer to this as *achievement-based inequality*.

At the same time, these societies wielded a simple weapon to keep members humble: shame. Canadian anthropologist Richard Lee's account of his life among the !Kung in the Kalahari Desert illustrates how this might have worked among our ancestors.[4] The following is a tribesman's description of how a successful hunter was expected to conduct himself:

'He must first sit down in silence until someone else comes up to his fire and asks, "What did you see today?" He replies quietly, "Ah, I'm no good for hunting. I saw nothing at all ... Maybe just a tiny one." Then I smile to myself because I now know he has killed something big.'[5]

Don't get me wrong – pride has been around for ages and so has greed. But for thousands of years, *Homo puppy* did everything it could to squash these tendencies. As a member of the !Kung put it: 'We refuse one who boasts, for someday his pride will make him kill somebody. So we always speak of his meat as worthless. This way we cool his heart and make him gentle.'[6]

Also taboo among hunter-gatherers was stockpiling and hoarding. For most of our history we didn't collect things, but friendships. This never failed to amaze European explorers, who expressed incredulity at the generosity of the peoples they encountered. 'When you ask for something they have, they never say no,' Columbus wrote in his log. 'To the contrary, they offer to share with anyone.'[7]

Of course there were always individuals who refused to abide by the fair-share ethos. But those who became too arrogant or greedy ran the risk of being exiled. And if that didn't work, there was one final remedy.

Take the following incident which occurred among the !Kung. The main figure here is /Twi, a tribe member who was growing increasingly unmanageable and had already killed two people. The group was fed up: 'They all fired on him with poison arrows till he looked like a porcupine. Then, after he was dead, all the women as well as the men approached his body and stabbed him with spears, symbolically sharing the responsibility for his death.'[8]

Anthropologists think interventions like this must have taken place occasionally in prehistory, when tribes made short work of members who developed a superiority complex. This was one of the ways we humans domesticated ourselves: aggressive personalities had fewer opportunities to reproduce, while more amiable types had more offspring.[9]

For most of human history, then, men and women were more or less equal. Contrary to our stereotype of the caveman as a chest-beating gorilla with a club and a short fuse, our male ancestors were probably not machos. More like proto-feminists.

Scientists suspect that equality between the sexes offered *Homo sapiens* a key advantage over other hominins like Neanderthals. Field studies show that in male-dominated societies men mostly hang out with brothers and male cousins. In societies where authority is shared with women, by contrast, people tend to have more diverse social networks.[10] And, as we saw in Chapter 3, having more friends ultimately makes you smarter.

Sexual equality was also manifest in parenting. Men in primitive societies spent more time with their children than many fathers do now.[11] Child-rearing was a responsibility shared by the whole tribe: infants were held by everybody and sometimes even breastfed by different women. 'Such early experiences,' notes one anthropologist, 'help explain why children in foraging societies tend to acquire working models of their world as a "giving place".'[12] Where modern-day parents warn their children not to talk to strangers, in prehistory we were raised on a diet of trust.

And one more thing. There are strong indications that hunter-gatherers were also pretty laid-back about their love lives. 'Serial monogamists' is how some biologists describe us. Take the Hadza in Tanzania, where the lifetime average is two or three partners, and women do the choosing.[13] Or take the mountain-dwelling Ache in Paraguay, where women average as many as twelve husbands in a lifetime.[14] This large network of potential fathers can come in handy, as they can all take part in child-rearing.[15]

When a seventeenth-century missionary warned a member of the Innu tribe (in what is now Canada) about the dangers of infidelity, he replied, 'Thou hast no sense. You French people love only your own children; but we all love all the children of our tribe.'[16]

2

The more I learned about how our ancestors lived, the more questions I had.

If it was true that we once inhabited a world of liberty and equality, why did we ever leave? And if nomadic foragers had no trouble removing domineering leaders, why can't we seem to get rid of them now?

The standard explanation is that modern society can no longer survive without them. States and multinationals need kings, presidents, and CEOs because, as geographer Jared Diamond puts it, 'large populations can't function without leaders who make the decisions'.[17] No doubt this theory is music to the ears of many managers and monarchs. And it sounds perfectly plausible, for how could you possibly build a temple, a pyramid, or a city without a puppet master pulling the strings?

And yet, history offers plenty of examples of societies that built temples and even whole cities from the ground up without rigid hierarchy. In 1995 archaeologists started excavating a massive temple complex in southern Turkey, whose beautiful carved pillars weigh more than twenty tons apiece. Think Stonehenge, but far more impressive. When the pillars were dated, researchers were astounded to learn that the complex was more than eleven thousand years old. That probably made it too early to have been built by any farming society (with kings or bureaucrats at the helm). And, indeed, search as they might, archaeologists could find no trace of agriculture. This gigantic structure had to be the work of nomadic foragers.[18]

Göbekli Tepe (translated as 'Potbelly Hill') turns out to be the oldest temple in the world and an example of what scholars call a *collective work event*. Thousands of people contributed, and pilgrims came from far and wide to lend a hand. Upon its completion, there was a big celebration with a feast of roast gazelle (archaeologists found thousands of gazelle bones). Monuments like this one were not built to stroke some chieftain's ego. Their purpose was to bring people together.[19]

To be fair, there are clues that individuals did occasionally rise to power in prehistory. A good example is the opulent grave discovered in 1955 at Sungir, 125 miles north of Moscow. It boasted bracelets carved from polished woolly mammoth tusk, a headdress fashioned from fox teeth and thousands of ivory beads, all 30,000 years old. Graves like this must have been the final resting places of princes and princesses of a kind, long before we were building pyramids or cathedrals.[20]

Even so, such excavations are few and far between, constituting no more than a handful of burial sites separated by hundreds of miles. Scientists now hypothesise that on those rare occasions when rulers did rise to power they were soon toppled.[21] For tens of thousands of years we had efficient ways of taking down anyone who put on airs. Humour. Mockery. Gossip. And if that didn't work, an arrow in the backside.

But then abruptly, that system stopped working. Suddenly rulers sat tight and managed to hang onto their power. Again the question is: why?

3

To understand where things went wrong, we have to go back 15,000 years, to the end of the last ice age. Up until then, the planet had been sparsely populated and people banded together to stave off the cold. Rather than a struggle for survival, it was a *snuggle* for survival, in which we kept each other warm.[22]

Then the climate changed, turning the area between the Nile in the west and the Tigris in the east into a land of milk and honey. Here, survival no longer depended on banding together against the elements. With food in such plentiful supply, it made sense to stay put. Huts and temples were built, towns and villages took shape and the population grew.[23]

More importantly, people's possessions grew.

What was it Rousseau had to say about this? 'The first man, who, after enclosing a piece of ground, took it into his head to say, "This is mine"' – that's where it all started to go wrong.

It couldn't have been easy to convince people that land or animals – or even other human beings – could now belong to someone. After all, foragers had shared just about everything.[24] And this new practice of ownership meant inequality started to grow. When someone died, their possessions even got passed on to the next generation. Once this kind of inheritance came into play, the gap between rich and poor opened wide.

What is fascinating is that it's at this juncture, after the end of the last ice age, that wars first break out. Just as we started settling down in one place, archaeological research has determined, we built the first military fortifications. This is also when the first cave paintings appeared that depict archers going at each other, and legions of skeletons from around this time have been found to bear clear traces of violent injury.[25]

THE CURSE OF CIVILISATION

How did it come to this? Scholars think there were at least two causes. One, we now had belongings to fight over, starting with land. And two, settled life made us more distrustful of strangers. Foraging nomads had a fairly laid-back membership policy: you crossed paths with new people all the time and could easily join up with another group.[26] Villagers, on the other hand, grew more focused on their own communities and their own possessions. *Homo puppy* went from cosmopolitan to xenophobe.

On those occasions that we did band together with strangers, one of the main reasons was, ironically, to make war. Clans began forming alliances to defend against other clans. Leaders emerged, likely charismatic figures who'd proved their mettle on the battlefield. Each new conflict further secured their position. In time these generals grew so wedded to their authority that they'd no longer give it up, not even in peacetime.

Usually the generals found themselves forcibly deposed. 'There must have been thousands of upstarts,' one historian notes, 'who failed to make the leap to a permanent kingship.'[27] But there were also times when intervention came too late, when a general had already drummed up enough followers to shield himself from the plebs. Societies dominated by this breed of ruler only became more fixated on war.

If we want to understand the phenomenon of 'war', we have to look at the people calling the shots. The generals and kings, presidents and advisers: these are the Leviathans who wage war, knowing it boosts their power and prestige.[28] Consider the Old Testament, where the Prophet Samuel warns the Israelites of the dangers of accepting a king. It is one of the most prescient – and sinister – passages in the Bible:

> These will be the ways of the king who will reign over you: he will take your sons and appoint them to his chariots and to be his horsemen and to run before his chariots. And he will appoint

for himself commanders of thousands and commanders of fifties, and some to plough his ground and to reap his harvest, and to make his implements of war and the equipment of his chariots. He will take your daughters to be perfumers and cooks and bakers. He will take the best of your fields and vineyards and olive orchards and give them to his servants. He will take the tenth of your grain and of your vineyards and give it to his officers and to his servants. He will take your male servants and female servants and the best of your young men and your donkeys, and put them to his work. He will take the tenth of your flocks, and you shall be his slaves.

The advent of settlements and private property had ushered in a new age in the history of humankind. The 1 per cent began oppressing the 99 per cent, and smooth talkers ascended from commanders to generals and from chieftains to kings. The days of liberty, equality and fraternity were over.

4

Reading about these recent archaeological discoveries, my thoughts returned to Jean-Jacques Rousseau. Self-proclaimed 'realistic' writers have all too often brushed him off as a naive romantic. But it was beginning to look like Rousseau might be the true realist after all.

The French philosopher rejected the notion of the advance of civilisation. He rejected the idea – still taught in schools today – that we started out as grunting cavemen, all bashing each other's brains in. That it was agriculture and private property that finally brought us peace, safety and prosperity. And that these gifts were eagerly embraced by our ancestors, who were tired of going hungry and fighting all the time.

Nothing could be farther from the truth, Rousseau believed. Only once we settled in one place did things begin to fall apart, he thought, and that's just what the archaeology now shows. Rousseau saw the invention of farming as one big fiasco, and for this, too, we now have abundant scientific evidence.

For one thing, anthropologists have discovered that hunter-gatherers led a fairly cushy life, with work weeks averaging twenty to thirty hours, tops. And why not? Nature provided everything they needed, leaving plenty of time to relax, hang out and hook up.

Farmers, by contrast, had to toil in the fields and working the soil left little time for leisure. No pain, no grain. Some theologists even suspect that the story of the Fall alludes to the shift to organised agriculture, as starkly characterised by Genesis 3: 'By the sweat of your brow you shall eat bread.'[29]

Settled life exacted an especially heavy toll on women. The rise of private property and farming brought the age of proto-feminism to an end. Sons stayed on the paternal plot to tend the land and livestock, which meant brides now had to be fetched for the family farm. Over centuries, marriageable daughters were reduced to little more than commodities, to be bartered like cows or sheep.[30]

In their new families, these brides were viewed with suspicion, and only after presenting them with a son did women gain a measure of acceptance. A legitimate son, that is. It's no accident that female virginity turned into an obsession. Where in prehistory women had been free to come and go as they pleased, now they were being covered up and tethered down. The patriarchy was born.

And things just kept getting worse. Rousseau was right again when he said that settled farmers were not as healthy as nomadic foragers. As nomads, we got plenty of exercise and enjoyed a varied diet rich in vitamins and fibre, but as farmers we began consuming a monotonous menu of grains for breakfast, lunch and dinner.[31]

We also began living in closer confines, and near our own waste. We domesticated animals such as cows and goats and started drinking their milk. This turned towns into giant Petri dishes for mutating bacteria and viruses.[32] 'In following the history of civil society,' Rousseau remarked, 'we shall be telling also that of human sickness.'[33]

Infectious diseases like measles, smallpox, tuberculosis, syphilis, malaria, cholera and plague were all unheard of until we traded our nomadic lifestyle for farming. So where did they come from? From our new domesticated pets – or, more specifically, their microbes. We get measles by way of cows, while flu comes from a microscopic *ménage à trois* between humans, pigs and ducks, with new strains still emerging all the time.

The same with sexually transmitted diseases. Virtually unknown in nomadic times, among pastoralists they began running rampant. Why? The reason is rather embarrassing. When humans began raising livestock, they also invented bestiality. Read: sex with animals. As the world grew increasingly uptight, the odd farmer covertly forced himself on his flock.[34]

And that's the second spark for the male obsession with female virginity. Apart from the matter of legitimate offspring, it was also a fear of STDs. Kings and emperors, who had entire harems at their disposal, went to great lengths to ensure their partners were 'pure'. Hence the idea, still upheld by millions today, that sex before marriage is a sin.

Famines, floods, epidemics – no sooner had humans settled down in one place than we found ourselves battling an endless cycle of disasters. A single failed harvest or deadly virus was enough to wipe out whole populations. For *Homo puppy*, this must have been a bewildering turn of events. Why was this happening? Who was behind it?

Scholars agree that people have probably always believed in gods and spirits.[35] But the deities of our nomadic ancestors were not all that interested in the lives of mere mortals, let alone in punishing their infractions. Nomadic religions would have more closely resembled that described by an American anthropologist who spent years living with the Hadza nomads in Tanzania:

> I think one can say the Hadza do have religion, certainly a cosmology anyway, but it bears little resemblance to what most of us in complex societies (with Christianity, Islam, Hinduism, etc.) think of as religion. There are no churches, preachers, leaders, or religious guardians, no idols or images of gods, no regular organized meetings, no religious morality, no belief in an afterlife – theirs is nothing like the major religions.[36]

The emergence of the first large settlements triggered a seismic shift in religious life. Seeking to explain the catastrophes suddenly befalling us, we began to believe in vengeful and omnipotent beings, in gods who were enraged because of something we'd done.

A whole clerical class was put in charge of figuring out why the gods were so angry. Had we eaten something forbidden? Said something wrong? Had an illicit thought?[37] For the first time in history, we developed a notion of sin. And we began looking to priests to prescribe how we should do penance. Sometimes it was enough to pray or complete a strict set of rituals, but often we had to sacrifice cherished possessions – food or animals or even people.

We see this with the Aztecs, who established a vast industry for human sacrifice in their capital at Tenochtitlan. When the conquistadors marched into the city in 1519 and entered its largest temple they were stunned to see huge racks and towers piled high with thousands of human skulls. The purpose of these human sacrifices, scholars now believe, was not only to appease the gods. 'The killing of captives, even in a ritual context,' one archaeologist has observed, 'is a strong political statement [...] it's a way to control your own population.'[38]

Reflecting on all this misery – the famines, the plagues, the oppression – it's hard not to ask: *why?* Why did we ever think it would be a good idea to settle in one place? Why did we exchange our nomadic life of leisure and good health for a life of toil and trouble as farmers?

Scholars have been able to piece together a fairly decent picture of what happened. The first settlements were probably just too tempting: finding ourselves in an earthly paradise where the trees hung heavy with fruit and untold gazelle and caribou grazed, it must have seemed crazy not to stay put.

With farming, it was much the same. There was no lightbulb moment when somebody shouted: 'Eureka! Let's start planting crops!' Though our ancestors had been aware for tens of thousands of years that you could plant things and harvest them, they also knew enough not to go down that road. 'Why should we plant,' exclaimed one !Kung tribesman to an anthropologist, 'when there are so many mongongo nuts in the world?'[39]

The most logical explanation is that we fell into a trap. That trap was the fertile floodplain between the Tigris and the Euphrates, where crops grew without much effort. There we could sow in soil enriched by a soft layer of nutrient-rich sediment left behind each year by the receding waters. With nature doing most of the work, even the work-shy *Homo puppy* was willing to give farming a go.[40]

What our ancestors couldn't have foreseen was how humankind would proliferate. As their settlements grew denser, the population of wild animals declined. To compensate, the amount of land under cultivation had to be extended to areas not blessed with fertile soil. Now farming was not nearly so effortless. We had to plough and sow from dawn to dusk. Not being built for this kind of work, our bodies developed all kinds of aches and pains. We had evolved to gather berries and chill out, and now our lives were filled with hard, heavy labour.

So why didn't we just go back to our freewheeling way of life? Because it was too late. Not only were there too many mouths to feed, but by this time we'd also lost the knack of foraging. And we couldn't just pack up and head for greener pastures, because we were hemmed in by neighbouring settlements, and they didn't welcome trespassers. We were trapped.

It didn't take long before the farmers outnumbered the foragers. Farming settlements could harvest more food per acre, which meant they could also raise larger armies. Nomadic tribes that stuck to their traditional way of life had to fend off invading colonists and their infectious illnesses. In the end, tribes that refused to bow down to a despot were beaten down by force.[41]

The outbreak of these first clashes signalled the start of the great race that would shape world history. Villages were conquered by towns, towns were annexed by cities and cities were swallowed up by provinces as societies all frantically scaled up to meet the inexorable demands of war. This culminated in the final catastrophic event so lamented by Rousseau.

The birth of the state.

5

Let's return for a moment to the picture Thomas Hobbes painted of the first humans to walk the earth. He believed that an unfettered life pitted our forebears in a 'war of all against all'. It only makes sense that we'd rush to embrace the first Leviathans (chieftains and kings) and the security they promised. Says Hobbes.

We now know that our nomadic ancestors were actually *fleeing* these despots. The first states – think Uruk in Mesopotamia or the Egypt of the pharaohs – were, without exception, slave states.[42] People didn't *choose* to live crammed together, but were corralled by regimes ever-hungry for new subjects, as their slaves kept dying of pox and plague. (It's no accident that the Old Testament paint cities in such a negative light. From the failed Tower of Babel to the destruction of Sodom and Gomorrah, God's judgement of sin-ridden cities is loud and clear.)

It's ironic at best. The very things we hold up today as 'milestones of civilization', such as the invention of money, the development of writing, or the birth of legal institutions, started out as instruments of oppression. Take the first coins: we didn't begin minting money because we thought it would make life easier, but because rulers wanted an efficient way to levy taxes.[43] Or think about the earliest written texts: these weren't books of romantic poetry, but long lists of outstanding debts.[44]

And those legal institutions? The legendary Code of Hammurabi, the first code of law, was filled with punishments for helping slaves to escape.[45] In ancient Athens, the cradle of western democracy, *two-thirds* of the population was enslaved. Great thinkers like Plato and Aristotle believed that, without slavery, civilisation could not exist.

Perhaps the best illustration of the true nature of states is the Great Wall of China, a wonder of the world meant to keep dangerous 'barbarians' out – but also to lock subjects in. Effectively it made the Chinese empire the largest open-air prison the world has ever known.[46]

And then there's that painful taboo in America's past on which most history textbooks are silent. One of the few willing to acknowledge it was Founding Father Benjamin Franklin. In the very same years that Rousseau was writing his books, Franklin admitted that 'No European who has tasted Savage Life can afterwards bear to live in our societies.'[47] He described how 'civilised' white men and women who were captured and subsequently released by Indians invariably would 'take the first good Opportunity of escaping again into the Woods'.

Colonists fled into the wilderness by the hundreds, whereas the reverse rarely happened.[48] And who could blame them? Living as Indians, they enjoyed more freedoms than they did as farmers and taxpayers. For women, the appeal was even greater. 'We could work as leisurely as we pleased,' said a colonial woman who hid from countrymen sent to 'rescue' her.[49] 'Here, I have no master,' another told a French diplomat. 'I shall marry if I wish and be unmarried again when I wish. Is there a single woman as independent as I in your cities?'[50]

In recent centuries, whole libraries have been written about the rise and fall of civilisations. Think about the overgrown pyramids of the Maya and the abandoned temples of the Greeks.[51] Underpinning all these books is the premise that, when civilisations fail, everything gets worse, plunging the world into 'dark ages'.

Modern scholars suggest it would be more accurate to characterise those dark ages as a reprieve, when the enslaved regained their freedom, infectious disease diminished, diet improved and culture flourished. In his brilliant book *Against the Grain* (2017), anthropologist James C. Scott points out that masterpieces like the *Iliad* and the *Odyssey* originated during the 'Greek Dark Ages' (1,110 to 700 BC) immediately following the collapse of Mycenaean civilisation. Not until much later would they be recorded by Homer.[52]

So why is our perception of 'barbarians' so negative? Why do we automatically equate a lack of 'civilisation' with dark times? History, as we know, is written by the victors. The earliest texts abound with propaganda for states and sovereigns, put out by oppressors seeking to elevate themselves while looking down on everybody else. The word 'barbarian' was itself coined as a catch-all for anyone who didn't speak ancient Greek.

That's how our sense of history gets flipped upside down. Civilisation has become synonymous with peace and progress, and wilderness with war and decline. In reality, for most of human existence, it was the other way around.

6

Thomas Hobbes, the old philosopher, could not have been more off the mark. He characterised the life and times of our ancestors as 'nasty, brutish and short', but a truer description would have been friendly, peaceful and healthy.

The irony is that the curse of civilisation dogged Hobbes throughout his life. Take the plague that killed his patron in 1628, and the looming civil war that forced him to flee England for Paris in 1640. The man's take on humanity was rooted in his own experience with disease and war, calamities which were virtually unknown for the first 95 per cent of human history. Hobbes has somehow gone down in history as the 'father of realism', yet his view of human nature is anything but realistic.

But is civilisation *all* bad? Hasn't it brought us many good things, too? Aside from war and greed, hasn't the modern world also given us much to be thankful for?

Of course it has. But it's easy to forget that genuine progress is a *very* recent phenomenon. Up until the French Revolution

(1789), almost all states everywhere were fuelled by forced labour. Until 1800, at least three-quarters of the global population lived in bondage to a wealthy lord.[53] More than 90 per cent of the population worked the land, and more than 80 per cent lived in dire poverty.[54] In the words of Rousseau: 'Man is born free and everywhere he is in chains.'[55]

For ages civilisation was a disaster. The advent of cities and states, of agriculture and writing, didn't bring prosperity to most people, but suffering. Only in the last two centuries – the blink of an eye – have things got better so quickly that we've forgotten how abysmal life used to be. If you take the history of civilisation and clock it over twenty-four hours, the first twenty-three hours and forty-five minutes would be sheer misery. Only in the final fifteen minutes would civil society start to look like a good idea.

In those final fifteen minutes we've stamped out most infectious diseases. Vaccines now save more lives *each year* than would have been spared if we'd had world peace for the entire twentieth century.[56] Second, we're now richer than ever before. The number of people living in extreme poverty has dropped to under 10 per cent.[57] And, third, slavery has been abolished.

In 1842, the British consul general wrote to the Sultan of Morocco to ask what he was doing to prohibit the slave trade. The Sultan was surprised: 'The traffic in slaves is a matter on which all sects and nations have agreed from the time of the sons of Adam.'[58] Little did he know that, 150 years later, slavery would be officially banned around the world.[59]

Last and best of all, we've entered the most peaceful age ever.[60] In the Middle Ages as much as 12 per cent of Europe's and Asia's populations died violent deaths. But in the last hundred years – including two world wars – this figure has plummeted to 1.3 per cent worldwide.[61] (In the US it's now 0.7 per cent and in the Netherlands, where I live, it's less than 0.1 per cent.)[62]

There's no reason to be fatalistic about civil society. We can choose to organise our cities and states in new ways that benefit everyone. The curse of civilisation can be lifted. Will we manage to do so? Can we survive and thrive in the long term? Nobody knows. There's no denying the progress of the last decades, but at the same time we're faced with an ecological crisis on an existential scale. The planet is warming, species are dying out and the pressing question now is: how sustainable is our civilised lifestyle?

I'm often reminded of what a Chinese politician said in the 1970s when asked about the effects of the French Revolution of 1789. 'It's a little too soon to say,' he allegedly responded.[63]

Maybe the same applies to civilisation. Is it a good idea?

Too soon to say.

6

The Mystery of Easter Island

By now, my whole understanding of human history had shifted. Modern science has made short work of the veneer theory of civilisation. We've amassed plenty of counter-evidence over the past couple decades, and it continues to pile up.

Admittedly, our knowledge about prehistory will never be watertight. We'll never solve all the riddles surrounding our ancestors' lives. Piecing together their archaeological puzzle involves a fair share of guesswork, and we should always be wary of projecting modern anthropological findings onto the past.

That's why I want to take one final look at what people do when left to their own devices. Suppose Mano and the other boys from the real-life *Lord of the Flies* weren't marooned alone. Suppose there'd been girls on the boat, too, they went on to have children and grandchildren, and 'Ata wasn't found until hundreds of years later.

What would have happened? What does society look like when it develops in isolation?

We can, of course, take what we've learned so far about prehistoric life and try to picture it. But there's no need to speculate when you can zoom in on a true, documented case study. On a remote island long obscured by myth and mystery, the insights of the previous chapters come together.

I

As a young man, Jacob Roggeveen made his father a promise: one day, he would find the Southern Land. Such a discovery would secure his place among history's exalted explorers and mean everlasting fame for his family.

It was thought to be situated somewhere in the Pacific Ocean. As a cartographer, Jacob's father Arent Roggeveen was convinced the continent had to exist to balance the land masses of the northern hemisphere. And then there were the stories brought back by travellers. The Portuguese navigator Pedro Fernandes de Queirós described the Southern Land as a paradise on earth, peopled by peaceable natives yearning for Christianity. It boasted fresh water, fertile soil and – minor detail – mountains of silver, gold and pearls.

It was 1 August 1721, forty years after his father's death, that Jacob finally set sail. Destination: the Southern Land. From his flagship the *Arend*, he commanded a fleet of three vessels, seventy cannon and a crew of 244. The sixty-two-year-old admiral had high hopes of making history. And he would, but little did he suspect how.

Jacob Roggeveen wouldn't establish a new civilisation. He'd discover an old one.[1]

What happened eight months later never ceases to amaze me. On Easter Sunday 1722, one of Roggeveen's vessels raised the flag. The *Arend* came alongside to find out what the crew had seen. The answer? Land. They'd spotted a small island off the starboard side.

The island had been formed hundreds of thousands of years earlier where three volcanoes converged. *Paasch Eyland*, as the Dutch crew christened it ('Easter Island'), spanned just over

Jacob Roggeveen's expedition in search of the Southern Land

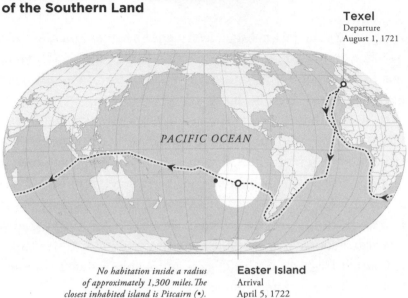

Texel
Departure
August 1, 1721

PACIFIC OCEAN

No habitation inside a radius of approximately 1,300 miles. The closest inhabited island is Pitcairn (•).

Easter Island
Arrival
April 5, 1722

a hundred square miles – a speck of land in the vast Pacific. The odds that Roggeveen would stumble upon it were more or less nil.

But the surprising existence of the island paled in comparison to their next discovery: there were people on this island.

As the Dutch approached, they saw a crowd gathered on the beach to meet them. Roggeveen was confounded. How had they got here? There wasn't a seaworthy boat in sight. Even more perplexing were the towering stone figures dotting the island – *moai*, the islanders called them – consisting of gigantic heads atop even bigger torsos, some thirty feet tall. 'We could not understand,' Roggeveen confided in his logbook, 'how it was possible that people who are destitute of heavy or thick timber, and also of stout cordage, out of which to construct gear, had been able to erect them.'[2]

When Roggeveen and his crew weighed anchor a week later, they had more questions than answers. Even today, this tiny island in the Pacific remains one of the most enigmatic places on earth, fuelling several centuries' worth of wild speculation. That the islanders were descended from the Inca, for example. That the statues had been put up by a race of twelve-foot-tall giants.[3] Or even that they were air-dropped by aliens (a Swiss hotel manager managed to sell seven million books on that theory).[4]

The truth is less fantastic – but not by much.

Thanks to DNA testing, we now know that explorers arrived long before Roggeveen came along. The Polynesians, those Vikings of the Pacific, found the island first.[5] With a courage verging on lunacy, they are thought to have set out from the Gambier Islands some sixteen hundred miles away, in open canoes, against the prevailing winds. How many such expeditions perished, we'll never know, but for this story only one needed to succeed.

And those colossal figures of the *moai*? When a young anthropologist named Katherine Routledge came to do fieldwork on the island in 1914, not a single statue remained standing. Instead they lay toppled, some broken and in pieces, overgrown with weeds.

How had this small society managed to make and move these monoliths? They lived on an island devoid of trees and didn't have the wheel at their disposal, much less cranes. Had the place been more populous once? Routledge put her questions to the island's oldest inhabitants. They told her stories of what had transpired here hundreds of years before. Chilling stories.[6]

Once upon a time, they said, two tribes lived on the island: the Long Ears and the Short Ears. They lived together in harmony until something happened that drove them apart, destroying the peace that had reigned for centuries and unleashing a bloody civil war. The Long Ears fled to the eastern part of the island and dug themselves in. The next morning, the Short Ears attacked the hideout from both sides and set it ablaze, incinerating the Long Ears in a trap of their own making. The remains of that trench are still visible today.

And that was only the beginning. In the years that followed, the situation degenerated into an all-out Hobbesian war, in which the Easter Islanders even resorted to eating each other. What had triggered all of this misery? Routledge could only guess. But clearly something must have happened to make a society destroy itself.

Years later, in 1955, a Norwegian adventurer by the name of Thor Heyerdahl mounted an expedition to Easter Island. Heyerdahl was something of a celebrity. A few years earlier he and five friends had cobbled together a raft and sailed it forty-three hundred miles from Peru to Polynesia, finally winding up wrecked on the island of Raroia. This long-distance rafting trip was proof for Heyerdahl that Polynesia had been populated by raft-rowing Incas. Though it failed to convince the experts, his theory did sell fifty million books.[7]

With the fortune he made on his bestseller, Heyerdahl was able to bankroll an expedition to Easter Island. He invited several eminent scientists to join him, among them William Mulloy, an American who would devote the rest of his life to studying Easter Island. 'I don't believe a damn thing you've published,' he assured Heyerdahl before they set out.[8]

Turns out scientist and daredevil got along surprisingly well, and not long after arriving on Easter Island the pair made a spectacular find. In the depths of a swamp, Heyerdahl's team discovered pollen from an unknown tree. They sent it to Stockholm for microscopic analysis by a leading palaeobotanist, who soon informed them of his conclusion. The island had once been home to a vast forest.

Slowly but surely the pieces began to come together. In 1974, a few years before his death, William Mulloy published the true story of Easter Island and the fate of its people.[9] Spoiler alert: it doesn't end well.

2

It all started with the mysterious *moai*.

For some reason, said Mulloy, the Easter Islanders couldn't get enough of these megaliths. One giant after another was chiselled from the rock and hauled into place. Jealous chieftains demanded larger and larger *moai*, more and more food was needed to feed the workforce and to transport the statues more and more of the island's trees were chopped down.

But a finite island cannot sustain infinite growth. There came a day when all the trees were gone. The soil eroded, causing crop yields to decline. Without wood for canoes, it was impossible to fish. Production of the statues stagnated and tensions grew. A war broke out between two tribes (the Long Ears and Short Ears that Katherine Routledge had been told about), culminating in a great battle around 1680 in which the Long Ears were almost entirely wiped out.

The surviving inhabitants then went on a destructive spree, Mulloy wrote, knocking down all the *moai*. Worse, they began to slake their hunger on one another. The islanders still tell the tale of their ancestors the cannibals, and a favourite insult is 'the flesh of your mother sticks between my teeth'.[10] Archaeologists have unearthed innumerable obsidian arrowheads, or *mata'a* – evidence of large-scale slaughter.

So when Jacob Roggeveen landed on Easter Island in 1722, he encountered a wretched population of just a few thousand individuals. Even today, the quarry at Rano Raraku where the *moai* were carved from the rock gives the impression of a workshop abandoned in sudden haste. Chisels lie where they were flung to the ground, with hundreds of *moai* left behind, unfinished.

William Mulloy's article represented a breakthrough in unravelling the mystery of Easter Island. Soon, a succession of other researchers were adding evidence to back up his case. Such as the two British geologists who in 1984 announced their discovery of fossil pollen grains in all three of the island's volcanic craters, confirming the hypothesis that the island had once been covered by a forest.[11]

Ultimately, it was world-famous geographer Jared Diamond who immortalised the tragic history of Easter Island.[12] In his 2005 bestseller *Collapse*, Diamond summed up the salient facts:

- Easter Island was populated by Polynesians early on, around the year 900.
- Analysis of the number of excavated dwellings indicates that the population once reached 15,000.
- The *moai* steadily increased in size, thus also increasing demand for manpower, food, and timber.
- The statues were transported horizontally on tree trunks, calling for a large workforce, lots of trees and a powerful leader to oversee operations.
- Eventually there were no more trees left, causing the soil to erode, agriculture to stagnate and famine among the inhabitants.
- Around 1680 a civil war broke out.

- When Jacob Roggeveen arrived in 1722, only a few thousand inhabitants remained. Innumerable *moai* had been knocked down and the islanders were eating one another.

The moral of this story?

The moral is about us. Set Easter Island and Planet Earth side by side and there are some disturbing parallels. Just consider: Easter Island is a speck in the vast ocean, the earth a speck in the vast cosmos. The islanders had no boats to flee; we have no rocket ships to take us away. Easter Island grew deforested and overpopulated; our planet is becoming polluted and overheated.

This leads us to a conclusion diametrically opposed to what I argued in the foregoing chapters. 'Humankind's covetousness is boundless,' archaeologists Paul Bahn and John Flenley write in their book *Easter Island, Earth Island*. 'Its selfishness appears to be genetically inborn.'[13]

Just when you thought you had cast off Hobbes's veneer theory, it doubles back like a boomerang.

The story of Easter Island seems to validate a cynical view of humankind. As our planet keeps warming and we keep on consuming and polluting, Easter Island looms as the perfect metaphor for our future. Forget *Homo puppy* and the noble savage. Our species seems more like a virus, or a cloud of locusts. A plague that spreads until everything is barren and broken – until it's too late.

So this is the lesson of Easter Island. Its calamitous history has been told and retold in documentaries and novels, in encyclopaedias and reports, in academic articles and popular science books. I've written about it myself. For a long time I believed the mystery of Easter Island had been solved by William Mulloy, Jared Diamond and their many cohorts. Because if so many leading experts draw identical, dismal conclusions, what's left to dispute?

Then I came across the work of Jan Boersema.

3

When I arrive at his office at Leiden University I can hear a Bach cantata playing in the background. At my knock, a man wearing a sharp flower-print shirt emerges from among the books.

Boersema may be an environmental biologist, but his shelves are also crammed with books on history and philosophy and his work draws on both the arts and the sciences. In 2002, this approach led him to make a simple yet profound discovery that contradicted all we thought we knew about Easter Island. He noticed something countless other researchers and writers had failed to see – or maybe just didn't want to see.

Boersema was preparing his inaugural lecture as a professor at the time and needed some background on Easter Island's decline. Wondering whether Roggeveen's logbook still existed, he went to check the library catalogue. Half an hour later, he had the *Journal of the Voyage of Discovery of Mr. Jacob Roggeveen* open on his desk.

'At first, I couldn't believe my eyes.' Boersema was expecting grisly scenes of carnage and cannibalism, but here in front of him was an upbeat travel log. 'There was nothing at all about a society in decline.'

Jacob Roggeveen characterised the Easter Islanders as friendly and healthy in appearance, with muscular physiques and gleaming white teeth. They didn't beg for food, they *offered* it to the Dutch crew. Roggeveen notes the island's 'exceptionally fertile' soil, but nowhere does he mention toppled statues, let alone weapons or cannibalism. Instead, he describes the island as an 'earthly paradise'.

'So then I wondered,' Boersema grins, 'what's going on here?'

Jan Boersema was one of the first scientists to express serious doubts about the widely accepted narrative of Easter Island's destruction. When I read his 2002 lecture, it dawned on me that the history of Easter Island is like a good mystery story: a scientific whodunit. So, like Boersema, let's try to unpack this mystery one step at a time. We'll verify the eyewitness accounts, check out the islanders' alibis, pin down the timeline as precisely as possible and zoom in on the murder weapons. We'll have to call on a whole gamut of disciplines during our investigation, from history to geology, and from anthropology to archaeology.[14]

Let's start by going back to the scene of the crime: the trench where the Long Ears hid and died in 1680. What's our source for this savage tale?

The first record we have are the memories Easter Islanders shared with Katherine Routledge in 1914. Now every investigator knows that human memory is fallible, and we're dealing here with memories passed down orally for generations. Imagine we had to explain what our ancestors were up to two or three hundred years ago. Then imagine we had no history books and could only rely on memories of stories of memories.

Conclusion? Maybe Routledge's notes are not the best source.

But hearsay wasn't the only evidence for the slaughter. One of the members of Thor Heyerdahl's expedition, archaeologist Carlyle Smith, began excavating around the trench reputed to be the site of the Long Ear massacre. He took two samples of charcoal and sent them off to be dated. One sample was narrowed down to the year 1676. For Smith, this clinched it. Since the date corresponded to when oral tradition situated the slaughter and burning of the Long Ears, he decided the story checked out.[15]

Although Smith later added some caveats to this interpretation, and although subsequent analyses re-dated the charcoal sample to anywhere between 1460 and 1817, and although no human remains were ever found at the site, and although geologists established that the trench had not been dug but was a natural feature of the landscape, the myth of the slaughter of 1680 persisted.[16] And it continued to be propagated by Heyerdahl, Mulloy and Diamond.

The case for an intertribal war gets weaker still when considered in light of the forensic evidence. The theory was that the islanders turned to cannibalism because they were starving. But more recent archaeological analysis of the skeletons of hundreds of inhabitants has determined that, in fact, Roggeveen's observations were right: the people living on Easter Island at the beginning of the eighteenth century were healthy and fit.[17] There's nothing to indicate they were going hungry.

What, then, about the clues pointing to mass violence?

A team of anthropologists at the Smithsonian Institution recently examined 469 skulls from Easter Island and found no evidence whatsoever to indicate large-scale warfare among the natives. Indeed, only two of the skulls bore traces of injury that, at least hypothetically, could have been inflicted using one of those infamous *mata'a* (the obsidian arrowheads).[18]

But scientists no longer believe the *mata'a* were weapons. More than likely, they served as common paring knives – like the piece of obsidian that one of Roggeveen's captains observed a native using to peel a banana. After examining four hundred *mata'a* in 2016, an American research team concluded they would have been useless as weapons: they were too dull.[19]

This is not to say the Easter Islanders didn't know how to make deadly weapons. But, as the team's leader drily remarked, 'they chose not to'.[20]

So the plot thickens. Because if they didn't murder each other, what happened to the thousands who once lived on the island? Where did they all go? Roggeveen tells us there were only a couple of thousand people living on the island when he visited, whereas at one time, according to Jared Diamond, there were as many as 15,000. What's their alibi?

Let's start by looking at the method Diamond used to arrive at this figure. First he gauged how many houses had once been on the island, based on archaeological remains. Next he guesstimated how many people lived in one house. Then to complete his calculations, he rounded up. Doesn't exactly sound like a foolproof formula.

We could make a much better estimate of the population if we could pin down the timeframe in which the drama played out. Easter Island was originally thought to have been populated around the year 900, or even as early as 300. But more recently, advanced technology has fixed this date substantially later, to roughly the year 1100.[21]

Using this later date, Jan Boersema has done a simple calculation. Let's say that about a hundred Polynesian seafarers landed on Easter Island in the year 1100. And say the population grew 0.5 per cent a year (the maximum achievable by pre-industrial societies). This means there could have been up to twenty-two hundred inhabitants by the time Roggeveen came ashore. This number tallies nicely with estimates recorded by European voyagers who stopped at the island in the eighteenth century.

Which means those thousands of Easter Islanders who supposedly tortured, killed and ate each other have an excellent alibi.

They never existed.

The next unsolved riddle is what happened to Easter Island's forests. If Jared Diamond, William Mulloy and a whole host of other scientists are to be believed, all the trees were chopped down by greedy inhabitants who wanted to put up as many *moai* as possible. A Canadian historian even goes so far as to diagnose 'mania' and 'ideological pathology'.[22]

But if you do the maths, you realise pretty quickly that this conclusion is a little rash. Boersema reckons that about fifteen trees were needed to roll each of the one thousand stone statues into place. That comes out to 15,000 trees, tops. So how many trees were there on the island? According to ecological research, millions – possibly even as many as sixteen million! [23]

Most of these statues never even left Rano Raraku, the quarry where they were carved. Yet rather than being 'abandoned' when the island was suddenly plunged into civil war, scientists now think they were left there intentionally to serve as 'guardians' of the quarry.[24]

In the end, 493 statues were rolled to another spot. That may sound like a lot, but don't forget that for hundreds of years the Easter Islanders had the place to themselves. At most, they only moved one or two statues a year. Why didn't they stop at a nice round dozen? Boersema suspects there is a simple explanation for this, too. Boredom. 'Living on an island like that, you basically had a lot of time on your hands,' he laughs. 'All that hacking and hauling helped to structure the day.'[25]

I think making the *moai* should really be seen as a *collective work event*, much like the construction of the temple complex at Göbekli Tepe more than ten thousand years ago (see Chapter 5). Or more recently on the island of Nias, west of Sumatra, where in the early twentieth century as many as 525 men were observed to drag a large stone statue on a wooden sled.[26]

No doubt endeavours like these could have been carried out more efficiently, but that wasn't the point. These were not prestige projects dreamed up by some megalomaniacal ruler. They were communal rituals that brought people together.

Let there be no misunderstanding: the Easter Islanders chopped down a good share of the trees. Not only to move the *moai*, but also to harvest the sap inside, to build canoes and to clear land for crops. Even so, when it comes to explaining the disappearance of the entire forest, there's a more likely culprit. Its name is *Rattus exulans*, aka the Polynesian rat.

These rodents were probably stowaways on the first boats to arrive, and with no natural predators on Easter Island they were free to feed and breed. In the lab, rats double in number every forty-seven days. That means that in just three years, a single pair of rats can produce seventeen *million* offspring.

This was the real ecological disaster on Easter Island. Biologists suspect these fast-proliferating rats fed on the seeds of trees, stunting the forest's growth.[27]

For the Easter Islanders, deforestation was not that big a deal, because every felled tree also freed up arable land. In a 2013 article, archaeologist Mara Mulrooney demonstrated that food production actually went *up* after the trees were gone, thanks to the islanders' use of savvy farming techniques like layering small stones to protect crops from wind and retain heat and moisture.[28]

Even if the population had reached 15,000, archaeologists say there still would have been plenty of food to go around. Mulrooney goes so far as to suggest that Easter Island perhaps 'should be the poster-child of how human ingenuity can result in success, rather than failure'.[29]

4

That success was to be short-lived.

The plague that would ultimately destroy Easter Island came not from within, but arrived on European ships. This tragic chapter opened on 7 April 1722, as Jacob Roggeveen and his crew were preparing to go ashore. A naked man came paddling up in a boat. Judged to be in his fifties, he was solidly built, had dark, tattooed skin and sported a goatee.

Once aboard, the fellow made an animated impression. He expressed amazement at 'the great height of the masts, the thickness of the ropes, the sails, the cannon, which he touched with great care, and also everything else that he saw'.[30] He had the fright of his life when he saw himself reflected in a mirror, when the ship's bell sounded, and when he poured a proffered glass of brandy into his eyes.

What impressed Roggeveen most was the islander's high spirits. He danced, he sang, he laughed and uttered repeated cries of 'O dorroga! O dorroga!' It wasn't until much later that scholars determined he was probably shouting 'Welcome'.

A bitter welcome it would be. Roggeveen moored with 134 men in three ships and two sloops. While the Easter Islanders showed every indication of delight, the Dutchmen lined up in battle formation. And then, without warning, four or five shots rang out. Someone shouted 'Now, now, open fire!' Thirty more shots followed. The islanders fled inland, leaving about ten dead on the beach. Among them was the friendly native who had originally greeted the fleet with 'O dorroga!'

Roggeveen was furious with the offenders, who claimed it had been a misunderstanding, but his journal makes no mention of punishment. When evening fell, Roggeveen insisted they leave, eager to resume his mission to find the Southern Land.

It would be forty-eight years before another fleet stopped at Easter Island. The expedition led by captain Don Felipe González planted three wooden crosses, raised the Spanish flag and claimed the island for the Virgin Mary. The Easter Islanders didn't seem to mind.

'There was not the least sign of animosity,' the conquerors noted.[31] When the Spaniards presented the inhabitants with a new bow and arrow, the peaceable natives were at a loss as to what to do with the gift. In the end, they opted to wear it like a necklace.

Four years later, in 1774, came an English expedition under the command of James Cook. It was Captain Cook who, after three epic voyages across the Pacific Ocean, would finally prove the Southern Land a myth. He joined the illustrious ranks of history's great explorers, while the name Roggeveen is long forgotten.

This is an engraving of a drawing by the artist Gaspard Duché de Vancy, who visited Easter Island on April 9, 1786. The image probably says more about this Frenchman and his colonial viewpoint than the natives of Easter Island. That it survived at all is something of a miracle, de Vancy having been part of an ill-fated expedition led by the explorer Jean-François de Galaup, Comte de La Pérouse. In 1787 the French arrived on the Kamchatka Peninsula in north-east Russia. There, just to be safe, La Pérouse decided to send home an advance report of his voyage (including this illustration). A year later his expedition was shipwrecked. What exactly befell La Pérouse, the expedition's artist De Vancy, and the rest of the crew is a mystery that scholars are still trying to unravel today. Source: Hulton Archive.

Cook's exalted stature may explain why the doomsayers put so much faith in his observations about Easter Island. Cook was the first to report on the toppled *moai* and – perhaps more importantly – described the natives as 'small, lean, timid, and miserable'.

Or, rather, that's what he's always quoted as writing. Oddly enough, when a University of Toronto researcher reread Cook's logbook, this unflattering description was nowhere to be found.[32] Instead, Cook reports that the inhabitants were 'brisk and active, have good features, and not disagreeable countenances; are friendly and hospitable to strangers'.[33]

So where did Cook make that scathing judgement? Where can we find this quote that fits so neatly with the narrative of Easter Island's collapse and even made it into the hallowed pages of the scientific journal *Nature*?[34] Jared Diamond cites as his sources Paul Bahn and John Flenley (authors of the book *Easter Island, Earth Island*), but they, in turn, cite none. I decided to try tracking down the mysterious quote myself. After a long day in the library I found it, in a dry book written for an academic readership in 1961.[35]

Subject? The Norwegian expedition to Easter Island. Author? Thor Heyerdahl.

That's right: the source of Cook's mangled quotation was none other than the Norwegian adventurer and champion of some rather hare-brained ideas. This is the same man who'd just published a popular bestseller in which he fantasised that the island was originally populated by long-eared Incas before being inundated by short-eared Polynesian cannibals. The same Thor Heyerdahl who recast Cook's 'harmless and friendly' islanders as a population of 'primitive cannibals'.[36]

This is how myths are born.

Meanwhile, there's one mystery still to be solved. Why did the Easter Islanders destroy their monumental statues?

For the answer, we have to go back to Jacob Roggeveen's journal. Until his arrival, the island's inhabitants had for hundreds of years supposed themselves to be completely alone in the world. It's probably no accident that all the *moai* faced inwards, towards the island, rather than outwards, towards the sea.

Then, after all that time, three gigantic ships appeared on the horizon. What would the islanders have thought of these strange Dutchmen, with their marvellous ships and their awful firepower? Were they prophets? Or gods? Their arrival and the massacre on the beach must have been a profound shock. 'Even their children's children in that place will in times to come be able to recount the story of it,' predicted one of the Dutch sailors.[37]

Next to come ashore with much pomp and fanfare were the Spaniards. They put on a ceremonial procession complete with drums and flag-waving and capped the show with three thunderous cannon shots.

Would it be a stretch to presume these events made an impact on the islanders and how they viewed the world? Where Roggeveen described seeing them kneel before the *moai*, Cook said the statues were no longer 'looked upon as idols by the present inhabitants, whatever they might have been in the days of the Dutch'. What's more, he noted, the islanders 'do not even repair the foundations of those which are going to decay'.[38]

By 1804, according to a Russian sailor's account, only a few of the *moai* were still standing. The rest had perhaps fallen over, or been knocked over intentionally, or maybe a little of both.[39] Whatever the case, the traditions surrounding the *moai* faded into obscurity and we'll never know precisely why. Two hypotheses have been put forward, either or both of which may be true. One is that the islanders found a new pastime. After the forests were gone, it got more difficult to move the megaliths around, so people devised new ways to fill their days.[40]

The other hypothesis involves what scholars call a 'cargo cult'. That is, an obsession with westerners and their stuff.[41] The Easter Islanders, for some reason, developed a fascination with hats. One French expedition lost all their headwear within a day of their arrival, causing great hilarity among the islanders.

It was also around this time that the island's inhabitants erected a house in the shape of a European ship, built stone mounds resembling boats and engaged in rituals mimicking European sailors. Scholars believe it may have been an attempt to will these foreigners to return with their strange and welcome gifts.

And return they did, but this time they didn't bring merchandise to trade. This time, the islanders were to become merchandise themselves.

5

The first slave ship appeared on the horizon one dark day in 1862.

Easter Island was the perfect prey for Peruvian slavers. It was isolated, home to a hale and hearty population and unclaimed by any world power. 'In brief,' sums up one historian, 'nobody was likely to know or care much about what happened to the people and the cost of removing them would be small.'[42]

At the final count, sixteen ships would sail off with a total of 1,407 people – a full third of the island's population. Some were tricked with false promises, others removed by force. It turns out the perpetrators were the very same slave traders who kidnapped the inhabitants of 'Ata (the island where the real-life *Lord of the Flies* would unfold a hundred years later). Once in Peru, the enslaved islanders started dropping like flies. Those who weren't worked to death in the mines succumbed to infectious diseases.

In 1863, the Peruvian government bowed to international pressure and agreed to ship the survivors back home. In preparation for their return, the islanders were gathered in the Peruvian port city of Callao. They got little to eat and, even worse, an American whaling ship berthed in the port had a crew member infected with smallpox. The virus spread. During the subsequent long sea voyage to Easter Island, corpses had to be thrown overboard daily, and in the end only fifteen of the 470 freed slaves made it home alive.

It would have been better for everyone if they'd died, too. Upon their return the virus spread among the rest of the population, sowing death and destruction. Easter Island's fate was sealed. Now Europeans who stopped at the island really did witness islanders turning against each other. There were heaps of bones and skulls, wrote one French sea captain, and the diseased were driven to such despair that dozens threw themselves off the cliffside to their deaths.

When the epidemic finally subsided in 1877, just 110 inhabitants remained — about the same number as had first paddled their canoes to the shore eight hundred years earlier. Traditions were lost, rituals forgotten, a culture decimated. The slavers and their diseases had finally accomplished what the native population and the rats had not. They destroyed Easter Island.

So what's left of the original story? Of that tale of self-centred islanders who ran their own civilisation into the ground?

Not much. There was no war, no famine, no eating of other people. Deforestation didn't make the land inhospitable, but more productive. There was no mass slaughter in or around 1680; the real decline didn't begin until centuries later, around 1860. And foreign visitors to the island didn't discover a dying civilisation — they pushed it off the cliff.

That's not to say the inhabitants didn't do some damage of their own, like accidentally introducing a plague of rats that wiped out indigenous plant and animal species. But after this rocky start, what stands out most is their resilience and adaptability. It turns out they were a lot smarter than the world long gave them credit for.

So is Easter Island still a fitting metaphor for our own future? A few days after my conversation with Professor Boersema, I saw a newspaper headline declaring: 'CLIMATE CHANGE ENDANGERS EASTER ISLAND STATUES.' Scientists have analysed the effects of rising ocean levels and coastal erosion, and this is the scenario they predict.[43]

I'm no sceptic when it comes to climate change. There's no doubt in my mind that this is the greatest challenge of our time – and that time is running out. What I am sceptical about, however, is the fatalistic rhetoric of collapse. Of the notion that we humans are inherently selfish, or worse, a plague upon the earth. I'm sceptical when this notion is peddled as 'realistic', and I'm sceptical when we're told there's no way out.

Too many environmental activists underestimate the resilience of humankind. My fear is that their cynicism can become a self-fulfilling prophecy – a nocebo that paralyses us with despair, while temperatures climb unabated. The climate movement, too, could use a new realism.

'There's a failure to recognise that not only problems but also solutions can grow exponentially,' Professor Boersema told me. 'There's no guarantee they will. But they can.'

For proof, we need only look to Easter Island. When the last tree was gone, the islanders reinvented farming, with new techniques to boost yields. The real story of Easter Island is the story of a resourceful and resilient people, of persistence in the face of long odds. It's not a tale of impending doom, but a wellspring of hope.

Part Two

AFTER AUSCHWITZ

'It's a wonder I haven't abandoned all my ideals, they seem so absurd and impractical. Yet I cling to them because I still believe, in spite of everything, that people are truly good at heart.'

Anne Frank (1929–45)

If it's true that human beings are kind-hearted by nature, then it's time to address the inevitable question. It's a question that made a number of German publishers less than enthusiastic about my book. And it's a question that continued to haunt me while I was writing it.

How do you explain Auschwitz?

How do you explain the raids and the pogroms, the genocide and concentration camps? Who were those willing executioners that signed on with Hitler? Or Stalin? With Mao? Or Pol Pot?

After the systematic murder of more than six million Jews, science and literature became obsessed with the question of how humans could be so cruel. It was tempting at first to see the Germans as a whole different animal, to chalk everything up to their twisted souls, sick minds, or barbaric culture. In any case, they were clearly nothing like us.

But there's a problem: the most heinous crime in human history wasn't committed in some primitive backwater. It happened in one of the richest, most advanced countries in the world — in the land of Kant and Goethe, of Beethoven and Bach.

Could it be that civil society was not a protective veneer after all? That Rousseau was right and civilisation an insidious rot? Around this time, a new scientific discipline rose to prominence and began to furnish disturbing proof that modern humans are indeed fundamentally flawed. That field was social psychology.

During the 1950s and 1960s, social psychologists began prying, probing and prodding to pin down what turns ordinary men and women into monsters. This new breed of scientist devised one experiment after another that showed humans are capable of appalling acts. A tweak in our situation is all it takes and — voila! — out comes the Nazi in each of us.

In the years that *Lord of the Flies* topped the bestseller lists, a young researcher named Stanley Milgram demonstrated how obediently people follow the orders even of dubious authority figures (Chapter 8), while the murder of a young woman in New York City laid the basis for hundreds of studies on apathy in the modern age (Chapter 9). And then there were the experiments by psychology professors Muzafer Sherif and Philip Zimbardo (Chapter 7), who demonstrated that good little boys can turn into camp tyrants at the drop of a hat.

What fascinates me is that all of these studies took place during a relatively short span of time. These were the wild west years of social psychology, when young hotshot researchers could soar to scientific stardom on the wings of shocking experiments.

Fifty years on, the young hotshots are dead and gone or travelling the globe as renowned professors. Their work is famous and continues to be taught to new generations of students. But now the archives of their post-war experiments have also been opened. For the first time, we can take a look behind the scenes.

7

In the Basement of Stanford University

I

It's 15 August 1971. Shortly before ten in the morning on the West Coast, Palo Alto police arrive in force to pull nine young men out of their beds. Five are booked for theft, four for armed robbery. Neighbours look on in surprise as the men are frisked, handcuffed and whisked away in the waiting police cars.

What the bystanders don't realise is that this is part of an experiment. An experiment that will go down in history as one of the most notorious scientific studies ever. An experiment that will make front-page news and become textbook fare for millions of college freshmen.

That same afternoon, the alleged criminals – in reality, innocent college students – descend the stone steps of Building 420 to the basement of the university's psychology department. A sign welcomes them to THE STANFORD COUNTY JAIL. At the bottom of the stairs waits another group of nine students, all dressed in uniforms, their eyes masked by mirrored sunglasses. Like the students in handcuffs, they're here to earn some extra cash. But these students won't be playing prisoner. They've been assigned the role of guard.

The prisoners are ordered to strip and are then lined up naked in the hallway. Chains are clapped around their ankles, nylon caps pulled down over their hair and each one gets a number by which he'll be addressed from this point on. Finally, they're given a smock to wear and locked behind bars, three to a cell.

What happens next will send shockwaves around the world. In a matter of days, the Stanford Prison Experiment spins out of control – and in the process reveals some grim truths about human nature.

The basement of Stanford University, August 1971.
Source: Philip G. Zimbardo.

It started with a group of ordinary, healthy young men. Several of them, when signing on for the study, called themselves pacifists.

By the second day things had already begun to unravel. A rebellion among the inmates was countered with fire extinguishers by the guards, and in the days that followed the guards devised all kinds of tactics to break their subordinates. In cells reeking of human faeces, the prisoners succumbed one by one to the effects of sleep deprivation and debasement, while the guards revelled in their power.

One inmate, prisoner 8612, went ballistic. Kicking his cell door, he screamed: 'I mean, Jesus Christ, I'm burning up inside! Don't you know? I want to get out! This is all fucked up inside! I can't stand another night! I just can't take it anymore!'[1]

The study's lead investigator, psychologist Philip Zimbardo, also got swept up in the drama. He played a prison superintendent determined to run a tight ship at any cost. Not until six days into the experiment did he finally call an end to the nightmare, after a horrified postgrad – his girlfriend – asked him what the hell he was doing. By then, five of the prisoners were exhibiting signs of 'extreme emotional depression, crying, rage and acute anxiety'.[2]

In the aftermath of the experiment, Zimbardo and his team were faced with a painful question: what happened? Today you can find the answer in just about any introductory psychology textbook. And in Hollywood blockbusters, Netflix documentaries and mega-bestsellers like Malcolm Gladwell's *The Tipping Point*. Or else swing by the office watercooler, where someone can probably fill you in.

The answer goes something like this. On 15 August 1971, a group of ordinary students morphed into monsters. Not because they were bad people, but because they'd been put in a bad situation. 'You can take normal people from good schools and happy families and good neighborhoods and powerfully affect their behavior,' Gladwell tells us, 'merely by changing the immediate details of their situation.'[3]

Philip Zimbardo would later swear up and down that nobody could have suspected his experiment would get so out of hand. Afterwards, he had to conclude that we're all capable of the most heinous acts. What happened in the basement of Stanford University had to be understood, he wrote, 'as a "natural" consequence of being in the uniform of a "guard" '.[4]

2

Few people know that, seventeen years earlier, another experiment was conducted which came to much the same conclusion. Largely forgotten outside academia, the Robbers Cave Experiment would inspire social psychologists for decades. And unlike the Stanford study, its subjects were not student volunteers, but unsuspecting children.

It's 19 June 1954. Twelve boys, all around eleven, are waiting at a bus stop in Oklahoma City. None of them know each other, but they're all from upstanding, churchgoing families. Their IQs are average, as are their grades at school. None are known troublemakers or get bullied. They're all well-adjusted, ordinary kids.

On this particular day, they're excited kids. That's because they're on their way to summer camp at Robbers Cave State Park in south-east Oklahoma. Famous as the one-time hideout of legendary outlaws like Belle Starr and Jesse James, the camp covers some two hundred acres of forest, lakes and caves. What the boys don't realise is that they'll be sharing this paradise with another group of campers that arrives the next day. And what they also don't know: this is a scientific experiment. The campers are the guinea pigs.

The study is in the hands of Turkish psychologist Muzafer Sherif, who has long been interested in how conflicts between

groups arise. His preparations for the camp have been meticulous and his instructions for the research team are clear: the boys
are to be free to do whatever they please, no holds barred.

In the first phase of the study, neither group of boys will be
aware of the other's existence. They'll stay in separate buildings
and assume they're alone in the park. Then, in the second week,
they'll be brought into careful contact. What will happen? Will
they become friends, or will all hell break loose?

The Robbers Cave Experiment is a story about well-behaved
little boys – 'the cream of the crop,' as Sherif later described
them – who in the space of a few days degenerate into 'wicked,
disturbed, and vicious bunches of youngsters'.[5] Sherif's camp
took place in the same year that William Golding published
his *Lord of the Flies*, but while Golding thought kids are bad by
nature, Sherif believed everything hinges on context.

Things start out pleasantly enough. During the first week,
when the two groups are still oblivious to one another's existence, the boys in each camp work together in perfect accord.
They build a rope bridge and a diving board. They grill
hamburgers and pitch tents. They run and play and they all
become fast friends.

The next week, the experiment takes a turn. The two groups,
having christened themselves the 'Rattlers' and the 'Eagles', are
cautiously introduced to one another. When the Rattlers hear
the Eagles playing on 'their' baseball field and challenge their
counterparts to a game, it touches off a week of rivalry and
competition. From there on out, things escalate quickly. On day
two the Eagles burn the Rattlers' flag after losing at tug-of-war.
The Rattlers retaliate with a midnight raid where they tear up
curtains and loot comic books. The Eagles decide to settle the
score by stuffing their socks with heavy rocks to use as weapons.
In the nick of time, the camp staff manage to intervene.

At the end of the week's tournament, the Eagles are declared the victors and get the coveted prize of shiny pocketknives. The Rattlers take revenge by mounting another raid and making off with all the prize booty. When confronted by the furious Eagles, the Rattlers only jeer. 'Come on, you yellow bellies,' taunts one of them, brandishing the knives.[6]

As the boys begin duking it out, Dr Sherif, posing as the camp caretaker, sits off to one side, busily scribbling his notes. He could tell already: this experiment was going to be a goldmine.

The story of the Robbers Cave Experiment has made a comeback in recent years, especially since Donald Trump was elected president of the United States. I can't tell you how many pundits have pointed to this study as the anecdotal key to understanding our times. Aren't the Rattlers and the Eagles a symbol for the ubiquitous clashes between left and right, conservative and progressive?

Television producers looked at the study's premise and saw a hit. In Holland, they attempted a thinly veiled remake aptly titled 'This Means War'. But shooting had to be terminated prematurely when it turned out the concept really did mean war.

Reasons enough to crack open Muzafer Sherif's original 1961 research report. Having read it, I can assure you: a page-turner it is not. On one of the first pages, Sherif tells us, 'Negative attitudes towards outgroups will be generated situationally.' Read: this means war.

But in among all the academic abstraction I found some interesting facts. For starters, it wasn't the kids themselves, but the experimenters who decided to hold a week of competitions. The Eagles weren't keen on the idea. 'Maybe we could make friends with those guys,' one boy suggested, 'and then somebody wouldn't get mad and have any grudges.'[7]

And at the researchers' insistence, the groups only played games that had clear-cut winners and losers, like baseball and tug-of-war. There were no consolation prizes, and the researchers manipulated scores to ensure the teams would stay in a neck-and-neck race.

Turns out these machinations were only the beginning.

3

I meet Gina Perry in Melbourne in the summer of 2017, just months before the publication of her book on the Robbers Cave Experiment. Perry is an Australian psychologist and was the first person to delve into the archives of Sherif's experiment. As she dug through reams of notes and recordings, she uncovered a story that contradicts everything the textbooks have been repeating for the past fifty years.

To begin with, Perry discovered that Sherif had tried to test his 'realistic conflict theory' before. He'd orchestrated another summer camp in 1953 outside the small town of Middle Grove in New York State. And there, too, he'd done his best to pit the boys against one another. The only thing Sherif was willing to say about it afterwards – tucked away in a footnote – was that the experiment had to be suspended 'due to various difficulties and unfavorable conditions'.[8]

In Melbourne, Perry tells me what she learned from the archives about what actually happened at that other, forgotten summer camp. Two days after their arrival, the boys had all become friends. They played games and ran wild in the woods, shot with bows and arrows and sang at the top of their lungs.

When day three rolled around, the experimenters split them up into two groups – the Panthers and the Pythons – and for the rest of the week they deployed every trick in the book to turn the two teams against each other. When the Panthers wanted to design team T-shirts that featured the olive branch of peace, the staff put a stop to it. A few days later, one of the experimenters tore down a Python tent, expecting the Panthers would take the heat for it. He looked on in frustration as the groups worked together to put the tent back up.

Next, the staff secretly raided the Panther camp, hoping the Pythons would get blamed. Once more, the boys helped each other out. One boy whose ukulele had been broken even called out the staff members and demanded an alibi. 'Maybe,' he accused, 'you just wanted to see what our reactions would be.'[9]

The mood within the research team soured as the week progressed. Their pricey experiment was on course to crash and burn. The boys weren't fighting like Sherif's 'realistic conflict theory' said they would, but instead remained the best of friends. Sherif blamed everyone but himself. He stayed up until two in the morning – pacing, as Perry could hear in the study's audio recordings – and drinking.

It was on one of the last evenings that tensions boiled over. While the campers lay peacefully asleep, Sherif threatened to punch a research assistant for not doing his best to sow discord among the children. The assistant grabbed a block of wood in self-defence. 'Dr. Sherif!' his voice echoed through the night, 'If you do it, I'm gonna hit you.'[10]

The children would eventually realise they were being manipulated, after one boy discovered a notebook containing detailed observations. After that, there was no choice but to call the experiment off. If anything had been proved, it was that once kids become friends it's very hard to turn them against each other. 'They misunderstood human nature,' said one participant about the psychologists, years later. 'They certainly misunderstood children.'[11]

4

If you think Dr Muzafer Sherif's manipulations are outrageous, they pale in comparison to the scenario cooked up seventeen years later. On the face of it, the Stanford Prison Experiment and the Robbers Cave Experiment have a lot in common. Both had twenty-four white, male subjects, and both were designed to prove that nice people can spontaneously turn evil.[12] But the Stanford Prison Experiment went a step further.

Philip Zimbardo's study wasn't just dubious. It was a hoax.

My own doubts surfaced on reading Zimbardo's book *The Lucifer Effect*, published in 2007. I had always assumed his prison 'guards' turned sadistic of their own accord. Zimbardo himself had claimed exactly that hundreds of times, in countless interviews, and in a hearing before the US Congress even testified that the guards 'made up their own rules for maintaining law, order, and respect'.[13]

But then, on page 55 of his book, Zimbardo suddenly mentions a meeting with the guards that took place on the Saturday preceding the experiment. That afternoon he briefed the guards on their role. There could be no mistaking his instructions:

> We can create a sense of frustration. We can create fear in them [...] We're going to take away their individuality in various ways. They're going to be wearing uniforms, and at no time will anybody call them by name; they will have numbers and be called only by their numbers. In general, what all this should create in them is a sense of powerlessness.[14]

When I came to this passage, I was stunned. Here was the supposedly independent scientist stating outright that he had drilled his guards. They hadn't come up with the idea to address the prisoners by numbers, or to wear sunglasses, or play sadistic games. It's what they were told to do.

Not only that, on the Saturday *before* the experiment started, Zimbardo was already talking about 'we' and 'they' as though he and the guards were on the same team. Which meant that the story he later told about losing himself in the role of prison superintendent as the experiment progressed couldn't be true. Zimbardo had been calling the shots from day one.

To grasp how fatal this is for objective research, it's important to know about what social scientists call *demand characteristics*. These are behaviours that subjects exhibit if they're able to guess at the aim of a study, thus turning a scientific experiment into a staged production. And in the Stanford Prison Experiment, as one research psychologist put it, 'the demands were everywhere'.[15]

What, then, did the guards themselves believe was expected of them? That they could sit around, maybe play some cards, and gossip about sports and girls? In a later interview, one student said he'd mapped out beforehand what he was going to do: 'I set out with a definite plan in mind, to try to force the action, force something to happen, so that the researchers would have something to work with. After all, what could they possibly learn from guys sitting around like it was a country club?'[16]

That the Stanford Prison Experiment hasn't been scrapped from the textbooks after confessions like this is bad enough. But it gets worse. In June 2013, French sociologist Thibault Le Texier stumbled across a TED Talk Zimbardo gave in 2009. As a part-time filmmaker, his attention was immediately caught by the images Zimbardo showed on screen. The raw footage of screaming students looked, to Le Texier's practised eye, like perfect material for a gripping documentary. So he decided to do some research.

Le Texier secured a grant from a French film fund and booked a flight to California. At Stanford he made two shocking discoveries. One was that he was the first to consult Zimbardo's archives. The other was what those archives contained. Le Texier's enthusiasm swiftly gave way to confusion and then to dismay: like Gina Perry, he found himself surrounded by piles of documents and recordings that presented what amounted to a whole different experiment.

'It took quite a while before I accepted the idea that it could all be fake,' Le Texier told me in the autumn of 2018, a year before his scathing analysis appeared in the world's leading academic psychology journal, *American Psychologist*. 'At first, I didn't want to believe it. I thought: no, this is a reputable professor at Stanford University. I must be wrong.'

But the evidence spoke for itself.

To begin with, it wasn't Zimbardo who dreamed up the experiment. It was one of his undergrads, a young man named David Jaffe. For a course assignment, he and four classmates thought it would be a neat idea to turn the basement of their dormitory into a jail. They drummed up a handful of willing friends and in May 1971 carried out their trial with six guards, six inmates and Jaffe himself as the warden.

The guards devised rules like 'Prisoners must address each other by number only' and 'Prisoners must always address the warden as "Mr Chief Correctional Officer."' In class the following Monday, Jaffe told all about his exciting 'experiment' and the intense emotions it had provoked in the participants. Zimbardo was sold. He had to try this out for himself.

There was only one aspect of the study that gave Zimbardo pause. Would he be able to find guards who were sadistic enough? Who could help him bring out the worst in people? The psychology professor decided to hire the undergrad as a consultant. 'I was asked to suggest tactics,' Jaffe subsequently explained, 'based on my previous experience as master sadist.'[17]

For forty years, in hundreds of interviews and articles, Philip Zimbardo steadfastly maintained that the guards in the Stanford Prison Experiment received no directives. That they'd thought it all up themselves: the rules, the punishments and the humiliations they inflicted on the prisoners. Zimbardo portrayed Jaffe as just another guard who – like the others – got swept up in the experiment.

Nothing could be further from the truth. Turns out eleven of the seventeen rules came from Jaffe. It was Jaffe who drafted a detailed protocol for the prisoners' arrival. Chaining them at the ankle? His idea. Undressing the inmates? That, too. Forcing them to stand around naked for fifteen minutes? Jaffe again.

On the Saturday before the experiment, Jaffe spent six hours with the other guards, explaining how they could use their chains and batons to best effect. 'I have a list of what happens,' he told them, 'some of the things that have to happen.'[18] After the whole ordeal was over, his fellow guards complimented him on his 'sado-creative ideas'.[19]

Meanwhile, Zimbardo was also contributing to the sadistic game plan. He drew up a tight schedule that would keep the inmates short on sleep, waking them up for roll calls at 2:30 a.m. and 6 a.m. He suggested push-ups as a good punishment for the prisoners, or putting thorny stickers or grass burrs in their blankets. And he thought solitary confinement might be a nice addition.

If you're wondering why Zimbardo took so much trouble to control the experiment, the answer is simple. Initially, Zimbardo wasn't interested in the guards. Initially, his experiment focused on the prisoners. He wanted to find out how prisoners would act under intense pressure. How bored would they get? How frustrated? How afraid?

The guards saw themselves as his research assistants, which makes sense considering that's precisely how Zimbardo treated them. Zimbardo's shocked response to their sadistic conduct, plus the idea that this was the true lesson of the experiment, were both manufactured after the fact. During the experiment, he and Jaffe pressured the guards to be extra tough on the inmates – then reprimanded those who failed to join in.

In an audio recording that has since surfaced, Jaffe can be heard taking this tack with 'soft' guard John Markus, pushing him as early as day two to take a harder line with the prisoners:

> Jaffe: 'Generally, you've been kind of in the background [...] we really want to get you active and involved because the guards have to know that every guard is going to be what we call a "tough guard" and so far, um ...'
>
> Markus: 'I'm not too tough ...'
>
> Jaffe: 'Yeah. Well, you have to try to get it in you.'
>
> Markus: 'I don't know about that ...'
>
> Jaffe: 'See, the thing is what I mean by tough is, you know, you have to be, um, *firm*, and you have to be in the action and, er, and that sort of thing. Um, it's really important for the workings of the experiment...'
>
> Markus: 'Excuse me, I'm sorry [...] if it was just entirely up to me, I wouldn't do anything. I would just let it cool off.'[20]

What's fascinating is that most guards in the Stanford Prison Experiment remained hesitant to apply 'tough' tactics at all, even under mounting pressure. Two-thirds refused to take part in the sadistic games. One-third treated the prisoners with kindness, to Zimbardo and his team's frustration. One of the guards resigned the Sunday before the experiment started, saying he couldn't go along with the instructions.

Most of the subjects stuck it out because Zimbardo paid well. They earned $15 a day – equivalent to about $100 now – but didn't get the money until afterwards. Guards and prisoners alike feared that if they didn't play along in Zimbardo's dramatic production, they wouldn't get paid.

But money was not enough incentive for one prisoner, who got so fed up after the first day that he wanted to quit. This was prisoner number 8612, twenty-two-year-old Douglas Korpi, who broke down on day two ('I mean, Jesus Christ [...] I just can't take it anymore!'[21]). His breakdown would feature in all the documentaries and become the most famous recording from the whole Stanford Prison Experiment.

A journalist looked him up in the summer of 2017.[22] Korpi told him the breakdown had been faked – play-acted from start to finish. Not that he'd ever made a secret of this. In fact, he told several people after the experiment ended: Zimbardo, for example, who ignored him, and a documentary filmmaker, who edited it out of his movie.

Douglas Korpi, who went on to earn a Ph.D. in psychology, later said that he initially enjoyed being in the experiment. The first day 'was really fun,' he recalled. 'I get to yell and scream and act all hysterical. I get to act like a prisoner. I was being a good employee. It was a great time.'[23]

The fun was short-lived. Korpi had signed up expecting to be able to spend time studying for exams, but once he was behind bars Zimbardo & Co. wouldn't let him have his textbooks. So the very next day he decided to call it quits.

To his surprise, Zimbardo refused to let him leave. Inmates would only be released if they exhibited physical or mental problems. So Korpi decided to fake it. First, he pretended to have a stomach ache. When that didn't work, he tried a mental breakdown ('I mean, Jesus Christ, I'm burning up inside! Don't you know? I want to get out! This is all fucked up inside! I can't stand another night! I just can't take it anymore!'[24]).

Those cries would become infamous the world over.

In the decades since the experiment, millions of people have fallen for Philip Zimbardo's staged farce.

'The worst thing,' one of the prisoners said in 2011, is that 'Zimbardo has been rewarded with a great deal of attention for forty years ...'[25] Zimbardo sent footage from the experiment to television stations before he'd even analysed his data. In the years that followed he would grow to be the most noted psychologist of his time, making it all the way to president of the American Psychological Association.[26]

In a 1990s documentary about the Stanford Prison Experiment, student guard Dave Eshelman wondered what might have happened if the researchers hadn't pushed the guards. 'We'll never know,' he sighed.[27]

It turns out that we would.

What Eshelman didn't know is that a pair of British psychologists were laying the groundwork for a second experiment. An experiment designed to answer the question: what happens to ordinary people when they don a uniform and step inside a prison?

5

The call from the BBC came in 2001.

It was the early days of reality TV. *Big Brother* had just debuted and television networks everywhere were busy brainstorming the next winning formula. So the BBC's request didn't come entirely out of the blue: would you be interested in taking another stab at that chilling experiment with the prisoners and the guards? But now for prime time?

For Alexander Haslam and Stephen Reicher, both doctors of psychology, it was a dream offer. The big problem with the Stanford Prison Experiment had always been that it was so unethical that no one dared to replicate it, and so Zimbardo had for decades enjoyed the last word. But now these two British psychologists were being offered an opportunity to do just that, on screen.

Haslam and Reicher said yes, on two conditions. One, they would have full control over the study. And two, an ethics committee would be authorised to halt the experiment at any time if things threatened to get out of hand.

In the months leading up to the broadcast, the British press was rife with speculation. To what depths would people sink? 'Is this reality TV gone mad?' wondered the *Guardian*.[28] Even Philip Zimbardo expressed disgust. 'Obviously they are doing the study in the hopes that high drama will be created as in my original study.'[29]

When the first episode of *The Experiment* aired on 1 May 2002, millions of people across Britain sat glued to their TVs. What happened next sent shockwaves around the ...

Well, actually, no it didn't.

What happened next was just about nothing. It took some real effort for me to sit through all four hour-long episodes. Rarely have I seen a programme this mind-numbingly dull.

Where did the BBC's formula go wrong? Haslam and Reicher left one thing out: they didn't tell the guards what to do. All the psychologists did was observe. They looked on from the sidelines as some ordinary guys sat around as though they were at the country club.

Things were just getting started when one guard announced he didn't feel suited to the role of guard: 'I'd rather be a prisoner, honestly ...'

On day two, another suggested sharing the guards' food with the prisoners to boost morale. Then on day four, when it looked like some sparks might fly, a guard advised a prisoner: 'If we can get to the end of this, we can go down the pub and have a drink.' Another guard chimed in, 'Let's discuss this like human beings.'

On day five, one of the prisoners proposed setting up a democracy. On day six, some prisoners escaped from their cells. They headed over to the guards' canteen to enjoy a smoke, where the guards soon came to join them. On day seven, the group voted in favour of creating a commune.

A couple of the guards did belatedly try to convince the group to go back to the original regime, but they weren't taken seriously. With the experiment at an impasse, the whole thing had to be called off. The final episode consists mostly of footage of the men lounging around on a sofa. At the very end, we're treated to some sentimental shots of the subjects hugging one another, and then one of the guard gives one of the prisoners his jacket.

Meanwhile, viewers are left feeling cheated. Where are the chained feet? Why no paper bags over heads? When can we expect the sadistic games to begin? The BBC broadcast four hours of end-to-end smoking, small talk and sitting around. Or, as the *Sunday Herald* summed it up, 'What happens when you put good men in an evil place and film it for telly? Erm, not that much actually.'[30]

For TV producers, the experiment exposed a harsh truth: if you leave ordinary people alone, nothing happens. Or worse, they'll try to start a pacifist commune.

From a scientific perspective, the experiment was a resounding success. Haslam and Reicher published more than ten articles about their results in prestigious academic journals. But for the rest, we can say that it was a failure. The BBC Prison Study has since faded into obscurity, while people still talk about the Stanford Prison Experiment.

And what does Philip Zimbardo have to say about all this? When a journalist asked him in 2018 if the new revelations about just how much was manipulated would change how people look at his experiment today, the psychologist responded that he didn't care. 'People can say whatever they want about it. It's the most famous study in the history of psychology at this point. There's no study that people talk about fifty years later. Ordinary people know about it. [...] It's got a life of its own now. [...] I'm not going to defend it anymore. The defense is its longevity.'[31]

8

Stanley Milgram and the Shock Machine

I

There's one psychological experiment even more famous than the Stanford Prison Experiment, and one psychologist who'd become more widely known than Philip Zimbardo. When I started to work on this book, I knew I couldn't ignore him.

Stanley Milgram.

Milgram was a young assistant professor when he launched his study on 18 June 1961. That day a full-page advertisement ran in the *New Haven Register*: 'We will pay you $4.00 for one hour of your time.'[1] The ad called for five hundred ordinary men – barbers and bartenders, builders and businessmen – to take part in research on human memory.

Hundreds of men visited Stanley Milgram's laboratory at Yale University over the next few months. Arriving in pairs, they would draw lots assigning one man to the role of 'teacher', the other to that of 'learner'. The teachers were seated in front of a large device which they were told was a shock machine. They were then instructed to perform a memory test with the learner, who was strapped to a chair in the next room. For every wrong answer, the teacher had to press a switch to administer an electric shock.

In reality, the learner was always a member of Milgram's team, and the machine didn't deliver shocks at all. But the teachers didn't know that. They thought this was a study on the effect of punishment on memory and didn't realise the study was really about them.

The shocks started small, a mere 15 volts. But each time the learner gave a wrong answer, a man in a grey lab coat directed the teacher to raise the voltage. From 15 volts to 30. From 30 volts to 45. And so on and so forth, no matter how loudly the learner in the next room screamed, and even after reaching the zone labelled 'DANGER: SEVERE SHOCK'. At 350 volts the learner pounded on the wall. After that, he went silent.

Milgram had asked some forty fellow psychologists to predict how far his test subjects would be willing to go. Unanimously, they said that at most 1 or 2 per cent — only downright psychopaths — would persist all the way to 450 volts.[2]

The real shock came after the experiment: 65 per cent of the study participants had continued right up to the furthest extreme and administered the full 450 volts. Apparently, two-thirds of those ordinary dads, pals and husbands were willing to electrocute a random stranger.[3]

Why? Because someone told them to.

Psychologist Stanley Milgram, twenty-eight years old at the time, became an instant celebrity. Just about every newspaper, radio station, and television channel covered his experiment. 'Sixty-five Percent in Test Blindly Obey Order to Inflict Pain', headlined the *New York Times*.[4] What kind of person, the paper asked, was capable of sending millions to the gas chambers? Judging from Milgram's findings, the answer was clear. All of us.

Stanley Milgram, who was Jewish, presented his research from the outset as the supreme explanation for the Holocaust. Where Muzafer Sherif hypothesised that war breaks out as soon as groups of people face off, and where Zimbardo (who went to school with Milgram) would claim we turn into monsters as soon as we don a uniform, Milgram's explanation was much more refined. More intelligent. And above all, more disturbing.

Stanley Milgram and his shock machine. Source: The Chronicle of Higher Education.

For Milgram, it all hinged on authority. Humans, he said, are creatures that will follow orders blindly. In his basement lab at Yale, grown men devolved into unthinking children, into trained Labradors that happily obeyed when commanded to 'sit', 'shake', or 'jump off a bridge'. It was all eerily similar to those Nazis who after the war continued to churn out the same old phrase: *Befehl ist Befehl* – an order's an order.

Milgram could draw only one conclusion: human nature comes with a fatal flaw programmed in – a defect that makes us act like obedient puppies and do the most appalling things.[5] 'If a system of death camps were set up in the United States,' claimed the psychologist, 'one would be able to find sufficient personnel for those camps in any medium-sized American town.'[6]

The timing of Milgram's experiment couldn't have been better. On the day the first volunteer walked into his lab, a controversial trial was entering its final week. Nazi war criminal Adolf Eichmann was being tried in Jerusalem before the eyes of seven hundred journalists. Among them was the Jewish philosopher Hannah Arendt, who was reporting on the case for the *New Yorker* magazine.

In pre-trial detention, Eichmann had undergone a psychological evaluation by six experts. None found symptoms of a behavioural disorder. The only weird thing about him, according to one of the doctors, was that he seemed 'more normal than normal'.[7] Eichmann, Arendt wrote, was neither psychopath nor monster. He was just as ordinary as all those barbers and bartenders, builders and businessmen who came into Milgram's lab. In the last sentence of her book, Arendt diagnosed the phenomenon: 'the banality of evil.'[8]

Milgram's study and Arendt's philosophy have been tied together since. Hannah Arendt would come to be regarded as one of the twentieth century's greatest philosophers; Stanley Milgram delivered the evidence to confirm her theory. A whole host of documentaries, novels, stage plays and television series were devoted to Milgram's notorious shock machine, which featured in everything from a movie with a young John Travolta, to an episode of *The Simpsons*, to a gameshow on French TV.

Fellow psychologist Muzafer Sherif even went so far as to say that 'Milgram's obedience experiment is the single greatest contribution to human knowledge ever made by the field of social psychology, perhaps psychology in general'.[9]

I'm going to be honest. Originally, I wanted to bring Milgram's experiments crashing down. When you're writing a book that champions the good in people, there are several big challengers on your list. William Golding and his dark imagination.

Richard Dawkins and his selfish genes. Jared Diamond and his demoralising tale of Easter Island. And, of course, Philip Zimbardo, the world's best-known living psychologist.

But topping my list was Stanley Milgram. I know of no other study as cynical, as depressing and at the same time as famous as his experiments at the shock machine. By the time I'd completed a few months' research, I reckoned I'd gathered enough ammunition to settle with his legacy. For starters, there are his personal archives, recently opened to the public. It turns out that they contain quite a bit of dirty laundry.

'When I heard that archival material was available,' Gina Perry told me during my visit to Melbourne, 'I was eager to look behind the scenes.' (This is the same Gina Perry who exposed the Robbers Cave Experiment as a fraud; see Chapter 7.) And so began what Perry called 'a process of disillusionment', culminating in a scathing book documenting her findings. What she uncovered had turned her from Milgram fan into fierce critic.

Let's first take a look at what Perry found. Again, it's the story of a driven psychologist chasing prestige and acclaim. A man who misled and manipulated to get the results he wanted. A man who deliberately inflicted serious distress on trusting people who only wanted to help.

2

The date is 25 May 1962. The final three days of the experiment have begun. Nearly one thousand volunteers have had their turn at his shock machine, when Milgram realises something's missing. Pictures.

A hidden camera is hurriedly installed to record participants' reactions. It's during these sessions that Milgram finds his star subject, a man whose name would become synonymous with the banality of evil. Or, rather, his pseudonym: Fred Prozi. If you've ever seen footage of Milgram's experiments, in one of hundreds of documentaries or in a clip on YouTube, then you've probably seen Prozi in action. And just like Zimbardo and prisoner 8612, it was the Fred Prozi recordings that made Milgram's message hit home.

We see a friendly-looking, heavyset man of around fifty who, with evident reluctance, does what he's told. 'But he might be dead in there!' he cries in distress – and then presses the next switch.[10] Watching the drama unfold, the viewer is hooked, both horrified and fascinated to see how far Prozi will go.

It makes for sensational television, and Milgram knew it. 'Brilliant,' he called Prozi's performance. He was thrilled with Prozi's 'complete abdication and excellent tension', and determined to cast this as his movie's leading character.[11] If you're thinking Milgram sounds more like a director than a scientist, you're not far off, for it was as a director that he really shone.

Anyone who deviated from his script was brought to heel by the application of intense pressure. The man in the grey lab coat – a biology teacher Milgram had hired named John Williams – would make as many as eight or nine attempts to get people to continue pressing higher switches. He even came to blows with one forty-six-year-old woman who turned the shock machine off. Williams turned it back on and demanded she continue.[12]

'The slavish obedience to authority,' writes Gina Perry, 'comes to sound much more like bullying and coercion when you listen to these recordings.'[13]

The key question is whether the experimental subjects believed they were administering real shocks at all. Shortly after the experiment, Milgram wrote that 'with few exceptions subjects were convinced of the reality of the experimental situation'.[14] Yet his archives are filled with statements from participants expressing doubt. Perhaps that's not very surprising when you consider how bizarre this situation must have seemed. Were people seriously expected to believe that someone was being tortured and killed under the watchful eye of scientists from a prestigious institution like Yale?

When the study was over, Milgram sent participants a questionnaire. One question was: how believable did you find the situation? Not until ten years later did he finally publish their answers, in the very last chapter of his book about the experiments. This is where we discover that only 56 per cent of his subjects believed they were actually inflicting pain on the learner. And that's not all. A never-published analysis by one of Milgram's assistants reveals that the majority of people called it quits if they did believe the shocks were real.[15]

So if nearly half the participants thought the setup was fake, where does that leave Milgram's research? Publicly, Milgram described his discoveries as revealing 'profound and disturbing truths of human nature'. Privately, he had his doubts. 'Whether all of this ballyhoo points to significant science or merely effective theater is an open question,' he wrote in his personal journal in June 1962. 'I am inclined to accept the latter interpretation.'[16]

When he published his results in 1963, Milgram's shock experiment met with abhorrence. 'Open-eyed torture', 'vile' and 'in line with the human experiments of the Nazis' were just a few ways the press characterised what he'd done.[17] The public outcry led to new ethical guidelines for experimental research.

All that time Milgram was keeping another secret. He chose not to inform some six hundred participants afterwards that the shocks in the experiment had not been real. Milgram was afraid the truth about his research would get out and he'd no longer be able to find test subjects. And so hundreds of people were left thinking they'd electrocuted another human being.

'I actually checked the death notices in the *New Haven Register* for at least two weeks after the experiment,' one said later, 'to see if I had been involved and a contributing factor in the death of the so-called learner.'[18]

3

In the first version of this chapter, I left it at that. My conclusion was that, like Philip Zimbardo's sadistic play-acting, Milgram's research had been a farce.

But in the months after meeting Gina Perry I was plagued by a nagging doubt. Could it be that I was just a little too keen to kick the shock machine to the curb? I thought back to Milgram's poll among almost forty colleagues, asking them to forecast how many subjects would go up to the full 450 volts. Every single one had predicted that only people who were genuinely crazy or disturbed would press that final switch.

One thing is certain: those experts were dead wrong. Even factoring in Milgram's biased point of view, his bullying assistant and the scepticism among his volunteers, there were still too many people who bowed to authority. Too many ordinary people believed the shocks were real and *still* continued to press the highest switch. No matter how you look at it, Milgram's results remain seriously disturbing.

And it's not only Milgram's results. Psychologists the world over have replicated his shock experiment in various iterations, with minor modifications (such as a shorter duration) to satisfy university ethics boards. As much as there is to criticise about these studies, the uncomfortable fact is that, over and over again, the outcome is the same.

Milgram's research seems unassailable. Bulletproof. Like a zombie that refuses to die, it just keeps coming back. 'People have tried to knock it down,' says one American psychologist, 'and it always comes up standing.'[19] Evidently, ordinary human beings are capable of terrible cruelty towards one another.

But why? Why does *Homo puppy* hit the 450-volt switch, if we're hardwired to be kind?

That's the question I needed to answer.

The first thing I wondered was whether Milgram's obedience experiments really tested obedience at all. Take the script he wrote up for Williams – the 'experimenter' in the grey lab coat – which directed him to give defiant subjects four specific 'prods'.

First: 'Please continue.'

Next: 'The experiment requires that you continue.'

After that: 'It is absolutely essential that you continue.'

And only in the last place: 'You have no other choice, you must go on.'

Modern-day psychologists have pointed out that only this last line is an order. And when you listen to the tapes, it's clear that as soon as Williams utters these words, everybody stops. The effect is instant *disobedience*. This was true in 1961, and it's been true when Milgram's experiment has been replicated since.[20]

Painstaking analyses of the hundreds of sessions at Milgram's shock machine furthermore reveal that subjects grew *more disobedient* the more overbearing the man in the grey coat became. Put differently: *Homo puppy* did not brainlessly follow the authority's orders. Turns out we have a downright aversion to bossy behaviour.

So then how was Milgram able to induce his subjects to keep pressing the switches? Alex Haslam and Steve Reicher, the psychologists behind the BBC Prison Study (see Chapter 7), have come up with an intriguing theory. Rather than submitting to the grey-coated experimenter, the participants decided to join him. Why? Because they trusted him.

Haslam and Reicher note that most people who volunteered for the study arrived feeling helpful. They wanted to help Mr Williams with his work. This would explain why the percentage of general goodwill declined when Milgram conducted the experiment in a plain office as opposed to the lofty setting of Yale. It could also explain why 'prods' invoking a scientific objective (like 'The experiment requires that you continue') were the most effective,[21] and why the participants behaved not like mindless robots, but were racked with doubt.

On the one hand, the teachers identified with the man in the grey lab coat, who kept repeating that the whole thing was in the interest of science. On the other, they couldn't ignore the suffering of the learner in the other room. Participants repeatedly cried, 'I can't take this anymore' and 'I'm quitting', even if they progressed to the next switch.

One man said afterwards that he had persisted for his daughter, a six-year-old with cerebral palsy. He hoped that the medical world would one day find a cure: 'I can only say that I was – look, I'm willing to do anything that's ah, to help humanity, let's put it that way.'[22]

When Milgram subsequently told his subjects that their contribution would benefit science, many expressed relief. 'I am happy to have been of service' was a typical response, and, 'Continue your experiments by all means as long as good can come of them. In this crazy mixed-up world of ours, every bit of goodness is needed.'[23]

When psychologist Don Mixon repeated Milgram's experiment in the seventies, he arrived at the same conclusion. He later noted, 'In fact, people go to great lengths, will suffer great distress, to be good. People got caught up in trying to be good ...'[24]

In other words, if you push people hard enough, if you poke and prod, bait and manipulate, many of us are indeed capable of doing evil. The road to hell is paved with good intentions. But evil doesn't live just beneath the surface; it takes immense effort to draw it out. And most importantly, evil has to be disguised as doing good.

Ironically, good intentions also played a major role in the Stanford Prison Experiment, from Chapter 7. Student guard Dave Eshelman, who wondered if he would have taken things as far if he hadn't been explicitly instructed to do so, also described himself as a 'scientist at heart'.[25] Afterwards, he said he felt he had done something positive, 'because I had contributed in some way to the understanding of human nature'.[26]

This was also true for David Jaffe, Zimbardo's assistant who came up with the original prison study concept. Jaffe encouraged the well-meaning guards to take a tougher line by pointing to the noble intentions behind the study. 'What we want to do,' he told a wavering guard, 'is be able to [...] go to the world with what we've done and say "Now look, this is what happens when you have Guards who behave this way." But in order to say that we have to have Guards who behave that way.'[27]

Ultimately, David Jaffe and Philip Zimbardo wanted their work to galvanise a complete overhaul of the prison system. 'Hopefully what will come out of this study is some very serious recommendations for reform,' Jaffe assured the guard. 'This is our goal. We're not trying to do this just because we're all, um, sadists.'[28]

4

That brings us back to Adolf Eichmann. On 11 April 1961, the Nazi officer's trial for war crimes began. Over the next fourteen weeks, hundreds of witnesses took the stand. For four-teen weeks the prosecution did its best to show what a monster Eichmann was.

But this was more than a court case alone. It was also a massive history lesson, a media spectacle to which millions of people tuned in. Among them was Stanley Milgram, described by his wife as a 'news addict', who closely followed the progress of the trial.[29]

Hannah Arendt, meanwhile, had a seat in the courtroom. 'The trouble with Eichmann,' she later wrote, 'was precisely that so many were like him, and that the many were neither perverted nor sadistic, that they were and still are, terribly and terrifyingly normal.'[30] In the years that followed, Eichmann came to stand for the mindless 'desk murderer' – for the banality of evil in each of us.

Only recently have historians come to some very different conclusions. When the Israeli secret service captured Eichmann in 1960, he'd been hiding out in Argentina. There, he'd been interviewed by former Dutch SS officer Willem Sassen for sev-eral months. Sassen hoped to get Eichmann to admit that the Holocaust was all a lie fabricated to discredit the Nazi regime. He was disappointed.

'I have no regrets!' Eichmann assured him.[31] Or as he'd already declared in 1945: 'I will leap into my grave laughing because the feeling that I have five million human beings on my conscience is for me a source of extraordinary satisfaction.'[32]

Reading through the thirteen hundred pages of interviews, teeming with warped ideas and fantasies, it's patently obvious that Eichmann was no brainless bureaucrat. He was a fanatic. He acted not out of indifference, but out of conviction. Like Milgram's experimental subjects, he did evil because he believed he was doing good.

Although transcripts of the Sassen interviews were available at the time of the trial, Eichmann managed to cast doubt on their authenticity. And so he put the whole world on the wrong track. All that time, the interview tapes lay mouldering in the Bundesarchiv in Koblenz, where the philosopher Bettina Stangneth found them fifty years later. What she heard confirmed that everything in Sassen's transcripts was true.

'I never did anything, great or small, without obtaining in advance express instructions from Adolf Hitler or any of my superiors,' Eichmann testified during the trial. This was a brazen lie. And his lie would be parroted by countless Nazis who professed that they were 'just following orders'.

Orders handed down within the Third Reich's bureaucratic machine tended to be vague, historians have since come to realise. Official commands were rarely issued, so Hitler's adherents had to rely on their own creativity. Rather than simply obeying their leader, historian Ian Kershaw explains that they 'worked towards him', attempting to act in the spirit of the Führer.[33] This inspired a culture of one-upmanship in which increasingly radical Nazis devised increasingly radical measures to get in Hitler's good graces.

In other words, the Holocaust wasn't the work of humans suddenly turned robots, just as Milgram's volunteers didn't press switches without stopping to think. The perpetrators believed they were on the right side of history. Auschwitz was the culmination of a long and complex historical process in which the voltage was upped step by step and evil was more convincingly passed off as good. The Nazi propaganda mill – with its writers and poets, its philosophers and politicians – had had years to do its work, blunting and poisoning the minds of the German people. *Homo puppy* was deceived and indoctrinated, brainwashed and manipulated.

Only then could the inconceivable happen.

Had Hannah Arendt been misled when she wrote that Eichmann wasn't a monster? Had she been taken in by his act on the stand?

That is the opinion of many historians, who cite her book as a case of 'great idea, bad example'.[34] But some philosophers disagree, arguing that these historians have failed to understand Arendt's thinking. For Arendt did in fact study parts of Sassen's interviews with Eichmann during the trial, and nowhere did she write that Eichmann was simply obeying orders.

What's more, Arendt was openly critical of Milgram's obedience experiments. As much as the young psychologist admired the philosopher, the sentiment wasn't mutual. Arendt accused Milgram of a 'naïve belief that temptation and coercion are really the same thing'.[35] And, unlike Milgram, she didn't think a Nazi was hiding in each of us.

Why *did* Milgram and Arendt enter the history books together? Some Arendt experts believe it's because she was misinterpreted. She was one of those philosophers who spoke in aphorisms, using enigmatic phraseology that could easily be misunderstood. Take her statement that Eichmann 'did not think'. She didn't say he was a robotic desk killer, but, rather, as Arendt expert Roger Berkowitz points out, that Eichmann was unable to think from someone else's perspective.[36]

In point of fact, Hannah Arendt was one of those rare philosophers who believe that most people, deep down, are decent.[37] She argued that our need for love and friendship is more human than any inclination towards hate and violence. And when we do choose the path of evil, we feel compelled to hide behind lies and clichés that give us a semblance of virtue.

Eichmann was a prime example. He'd convinced himself he'd done a great deed, something historic for which he'd be admired by future generations. That didn't make him a monster or a robot. It made him a joiner. Many years later, psychologists would reach the same conclusion about Milgram's research: the shock experiments were not about obedience. They were about conformity.

It's astonishing how far ahead of her time Hannah Arendt was when she made precisely the same observation.

Sadly, Stanley Milgram's simplistic deductions (that humans submit to evil without thinking) made a more lasting impression than Hannah Arendt's layered philosophy (that humans are tempted by evil masquerading as good). This speaks to Milgram's directorial talent, to his eye for drama and his astute sense of what works on television.

But above all, I think what made Milgram famous was that he furnished evidence to support an age-old belief. 'The experiments seemed to offer strong support,' writes psychologist Don Mixon, 'for history's oldest, most momentous self-fulfilling prophecy – that we are born sinners. Most people, even atheists, believe that it is good for us to be reminded of our sinful nature.'[38]

What makes us so eager to believe in our own corruption? Why does veneer theory keep returning in so many permutations? I suspect it has a lot to do with convenience. In a weird way, to believe in our own sinful nature is comforting. It provides a kind of absolution. Because if most people are bad, then engagement and resistance aren't worth the effort.

Belief in humankind's sinful nature also provides a tidy explanation for the existence of evil. When confronted with hatred or selfishness, you can tell yourself, 'Oh, well, that's just human nature.' But if you believe that people are essentially good, you have to question why evil exists at all. It implies that engagement and resistance are worthwhile, and it imposes an obligation to act.

In 2015, psychologist Matthew Hollander reviewed the taped recordings of 117 sessions at Milgram's shock machine.[39] After extensive analysis, he discovered a pattern. The subjects who managed to halt the experiment used three tactics:

1. Talk to the victim.
2. Remind the man in the grey lab coat of his responsibility.
3. Repeatedly refuse to continue.

Communication and confrontation, compassion and resistance. Hollander discovered that virtually all participants used these tactics – virtually all wanted to stop, after all – but that those who succeeded used them much more. The good news is: these are trainable skills. Resistance just takes practice. 'What distinguishes Milgram's heroes,' Hollander observes, 'is largely a teachable competency at resisting questionable authority.'[40]

If you think resistance is doomed to fail, then I have one last story for you on the subject. It takes place in Denmark during the Second World War. It's a story of ordinary people who demonstrated extraordinary courage. And it shows that resistance is always worthwhile, even when all seems lost.

5

The date is 28 September 1943.

In the headquarters of the Workers Assembly Building on 24 Rømersgade in Copenhagen, the Social Democratic Party leaders have all convened. A visitor in a Nazi uniform stands before them. They are staring at him in shock.

'The disaster is at hand,' the man is saying. 'Everything is planned in detail. Ships will anchor at the mooring off Copenhagen. Those of your poor Jewish countrymen who get caught by the Gestapo will forcibly be brought on board the ships and transported to an unknown fate.'[41]

The speaker is trembling and pale. Georg Ferdinand Duckwitz is his name. He will go down in history as 'the converted Nazi', and his warning will work a miracle.

The raid was set to take place on Friday 1 October 1943, following detailed plans drawn up by the SS. At the stroke of 8 p.m., hundreds of German troops would begin knocking on doors up and down the country to round up all the Danish Jews. They would be taken to the harbour and boarded onto a ship equipped to hold six thousand prisoners.

To put it in terms of the shock experiments: Denmark didn't go from 15 volts to 30 and from 30 volts to 45. The Danes would be told to give the highest 450-volt shock at once. Up until this moment there had been no discriminatory laws, no mandatory yellow badges, no confiscation of Jewish property. Danish Jews would find themselves being deported to Polish concentration camps before they knew what had hit them.

That, at least, was the plan.

On the appointed night, tens of thousands of ordinary Danes – barbers and bartenders, builders and businessmen – refused to press that last switch on the shock machine. That night, the Germans discovered that the Jews had been forewarned of the raid and that most had already fled. In fact, thanks to that warning, almost 99 per cent of Denmark's Jews survived the war.

How can we explain the miracle of Denmark? What made this country a beacon of light in a sea of darkness?

After the war, historians suggested a number of answers. One important factor was that the Nazis had not fully seized power in Denmark, wishing to preserve the impression that their two governments were working together in harmony. As a consequence, resistance against the Germans wasn't as risky in Denmark as in other countries, such as occupied Holland.

But ultimately one explanation stands out. 'The answer is undeniable,' writes historian Bo Lidegaard. 'The Danish Jews were protected by their compatriots' consistent engagement.'[42]

When news of the raid spread, resistance sprang up from every quarter. From churches, universities and the business community, from the royal family, the Lawyers Council and the Danish Women's National Council – all voiced their objection. Almost immediately, a network of escape routes was organised, even with no centralised planning and no attempt to coordinate the hundreds of individual efforts. There simply wasn't time. Thousands of Danes, rich and poor, young and old, understood that now was the time to act, and that to look away would be a betrayal of their country.

'Even where the request came from the Jews themselves,' historian Leni Yahil noted, 'these were never refused.'[43] Schools and hospitals threw open their doors. Small fishing villages took in hundreds of refugees. The Danish police also assisted where they could and refused to cooperate with the Nazis. 'We Danes don't barter with our Constitution,' stormed *Dansk Maanedspost*, a resistance newspaper, 'and least of all in the matter of citizens' equality.'[44]

Where mighty Germany was doped up on years of racist propaganda, modest Denmark was steeped in humanist spirit. Danish leaders had always insisted on the sanctity of the democratic rule of law. Anybody who sought to pit people against each other was not considered worthy to be called a Dane. There could be no such thing as a 'Jewish question'. There were only countrymen.

In a few short days, more than seven thousand Danish Jews were ferried in small fishing boats across the Sound separating Denmark from Sweden. Their rescue was a small but radiant point of light in a time of utter darkness. It was a triumph of humanity and courage. 'The Danish exception shows that the mobilization of civil society's humanism [...] is not only a theoretical possibility,' writes Lidegaard. 'It can be done. We know because it happened.'[45]

The Danish resistance turned out to be so contagious that even Hitler's most loyal followers in Denmark began to experience doubts. It became increasingly difficult for them to act as if they were backing a just cause. 'Even injustice needs a semblance of law,' Lidegaard observes. 'That is hard to find when the entire society denies the right of the stronger.'[46]

Only in Bulgaria and Italy did the Nazis encounter comparable resistance, and there the Jewish death toll was analogously low. Historians emphasise that the scale of deportations in occupied regions hinged on the extent of each country's collaboration.[47] In Denmark, Adolf Eichmann would tell Willem Sassen years later, the Germans had more difficulties than elsewhere. 'The result was meager ... I also had to recall my transports – it was for me a mighty disgrace.'[48]

To be clear, the Germans stationed in Denmark were no softies – as attested by the highest ranking Nazi there, Werner Best, better known as 'the Bloodhound of Paris'. Even Duckwitz, the converted Nazi in Copenhagen, had been a rabid anti-Semite throughout the 1930s. But as the years progressed, he became infected by the Danish spirit of humanity.

In her book *Eichmann in Jerusalem*, Hannah Arendt makes a fascinating observation about the rescue of the Danish Jews. 'It is the only case we know,' she wrote, 'in which the Nazis met with open native resistance, and the result seems to have been that those exposed to it changed their minds. They themselves apparently no longer looked upon the extermination of a whole people as a matter of course. They had met resistance based on principle, and their "toughness" had melted like butter in the sun ...'[49]

9

The Death of Catherine
Susan Genovese

I

There's one more story from the 1960s that needs to be told. Another story that exposes a painful truth about human nature. This time it's not about the things we do, but the things we fail to do. It's also a story that echoes what so many Germans, Dutch, French, Austrians and others across Europe would claim after millions of Jews were arrested, deported, and murdered in the Second World War.

Wir haben es nicht gewußt. 'We had no idea.'

It is 13 March 1964, a quarter past three in the morning. Catherine Susan Genovese drives her red Fiat past the NO PARKING sign just visible in the darkness and pulls up outside the Austin Street subway station.

Kitty, as everyone knows her, is a whirlwind of energy. Twenty-eight years old, she's crazy about dancing and has more friends than free time. Kitty loves New York City, and the city loves her. It's the place where she can be herself – the place she's free.

But that night it's cold outside, and Kitty's in a hurry to get home to her girlfriend. It's their first anniversary, and all Kitty wants to do is cuddle up with Mary Ann. Quickly switching off her lights and locking the car doors, she heads off towards their small apartment, less than a hundred feet away.

What Kitty doesn't know is that this will be the final hour of her life.

'Oh my God, he stabbed me! Help me!'

It's 3:19 a.m. The screams pierce the night, loud enough to wake the neighbourhood. In several apartments, lights flick on. Windows are raised and voices murmur in the night. One calls out, 'Let that girl alone.'

But Kitty's attacker returns. For the second time, he stabs her with his knife. Stumbling around the corner, she cries out 'I'm dying! I'm dying!'

Nobody comes outside. Nobody lifts a finger to help. Instead, dozens of neighbours peer through their windows, as though watching a reality show. One couple pulls up some chairs and dims the lights to get a clearer view.

When the attacker returns for a third time, he finds her lying at the foot of a stairwell just inside her apartment building. Upstairs, Mary Ann sleeps on, unaware.

Kitty's attacker stabs her again and again.

It's 3:50 a.m. when the first call comes into the police station. The caller is a neighbour who spent a long time deliberating what to do. Officers arrive on the scene within two minutes, but it's too late. 'I didn't want to get involved,' the caller admits to the police.[1]

These six words – 'I didn't want to get involved' – reverberated around the globe.

Initially, Kitty's death was one of the 636 murders committed in New York City that year.[2] A life cut short, a love lost, and the city moved on. But two weeks later, the story made the papers, and in time Kitty's murder would make it into the history books. Not because of the killer or the victim, but because of the spectators.

The media storm started on Good Friday – 27 March 1964. '37 Who Saw Murder Didn't Call the Police,' read the front page of the *New York Times*. The article opened with the following lines: 'For more than half an hour 38 respectable, law-abiding citizens in Queens watched a killer stalk and stab a woman in three separate attacks in Kew Gardens.' Kitty could still have been alive, the story said. As one detective put it, 'A phone call would have done it.'[3]

From Great Britain to Russia and from Japan to Iran, Kitty became big news. Here was proof, reported Soviet newspaper *Izvestia*, of capitalism's 'jungle morals'.[4] American society had become 'as sick as the one that crucified Jesus' preached a Brooklyn minister, while one columnist condemned his countrymen as 'a callous, chickenhearted and immoral people'.[5]

Journalists, photographers and TV crews swarmed Kew Gardens, where Kitty had lived. None of them could believe what a nice, neat, respectable neighbourhood it was. How could residents of a place like this display such complete and horrifying apathy?

It was the dulling effect of television, claimed one. No, said another, it was feminism that had turned men into wimps. Others thought it typified the anonymity of big-city life. And wasn't it reminiscent of the Germans after the Holocaust? They, too, had claimed ignorance: *We had no idea.*

But most widely accepted was the analysis furnished by Abe Rosenthal, metropolitan editor at the *New York Times* and a leading journalist of his generation. 'What happened in the apartments and houses on Austin Street,' he wrote, 'was a symptom of a terrible reality in the human condition.'[6]

When it comes down to it, we're alone.

This is the most famous picture of Kitty Genovese. It is a mug shot taken by the police in 1961, shortly after she was arrested for a misdemeanor (she worked at a bar and booked patrons' bets on horse races). Kitty was fined fifty dollars. The mug shot was cropped by the New York Times *and transmitted around the globe.*
Source: Wikimedia.

2

I was a student when I first read about Kitty Genovese. Like millions of people, I devoured journalist Malcolm Gladwell's debut book *The Tipping Point*, and it was on page 27 that I learned about those thirty-eight eyewitnesses.[7]

The story grabbed me, just as the stories about Milgram's shock machine and Zimbardo's prison had. 'I still get mail about it,' Rosenthal said years later. '[People] are obsessed by this story. It's like a jewel – you keep looking at it, and different things occur to you.'[8]

That fateful Friday the 13th became the subject of plays and songs. Entire episodes of *Seinfeld*, *Girls* and *Law and Order* were devoted to it. During a 1994 speech in Kew Gardens, President Bill Clinton recalled the 'chilling message' of Kitty's murder, and a US Deputy Secretary of Defense, Paul Wolfowitz, even used it as an oblique justification for the 2003 invasion of Iraq. (He suggested that Americans who opposed the war were just as apathetic as those thirty-eight witnesses.)[9]

The moral of this story seemed clear to me, too. Why didn't anybody come to Kitty Genovese's aid? Well, because people are callous and indifferent. This message was already gaining traction in the period that Kitty Genovese became a household name – it was the same era that *Lord of the Flies* became a bestseller, Adolf Eichmann stood trial, Stanley Milgram send shockwaves around the world and Philip Zimbardo launched his career.

But when I began reading up on research into the circumstances surrounding Kitty's death, I found myself on the trail of a whole different story. Again.

Bibb Latané and John Darley were two young psychologists at the time. They'd been studying what bystanders do in emergencies and noticed something strange. Not long after Kitty's murder, they decided to try an experiment. Their subjects were unsuspecting college students, who were asked to sit alone in a closed room and chat about college life with some of their peers over an intercom.

Except there were no other students: the researchers instead played a pre-recorded audio tape. 'I could really-er-use some help,' moaned a voice at some point, 'so if somebody would-er-give me a little h-help-uh-er-er-er-er-er c-could somebody-er-er-help-er-uh-uh-uh [choking sounds] ... I'm gonna die ...'[10]

What happened next? When a trial subject thought that they alone heard the cries for help, they rushed out into the corridor.

All of them, without exception, ran to intervene. But among those who were led to believe five other students were sitting in rooms nearby, only 62 per cent took action.[11] Voila: the bystander effect.

Latané and Darley's findings would be among the most pivotal contributions made to social psychology. Over the next twenty years, more than a thousand articles and books were published on how bystanders behave in emergencies.[12] Their results also explained the inaction of those thirty-eight witnesses in Kew Gardens: Kitty Genovese was dead not *in spite* of waking up the whole neighbourhood with her screams, but *because* of it.

This was exemplified by what one building resident later told a reporter. When her husband went to call the police, she held him back: 'I told him there must have been thirty calls already.'[13] Had Kitty been attacked in a deserted alleyway, with only one witness, she might have survived.

All this only fuelled Kitty's fame. Her story found its way into the top ten psychology textbooks and continues to be invoked by journalists and pundits to this day.[14] It's become nothing less than a modern parable on the perilous anonymity of big-city life.

3

For years I assumed the bystander effect was just an inevitable part of life in a metropolis. But then something happened in the very city where I work — something that forced me to reassess my assumptions.

It's 9 February 2016. At a quarter to four in the afternoon Sanne parks her white Alfa Romeo on Sloterkade, a canal-side street in Amsterdam.[15] She gets out and heads to the passenger side to take her toddler out of the car seat when, suddenly, she becomes aware the car is still rolling. Sanne barely manages to jump back behind the wheel, but it's too late for brakes. The car tips down into the canal and begins to sink.

The bad news: dozens of bystanders saw it happen.

No doubt even more people heard Sanne's screams. Just as in Kew Gardens, there are apartments overlooking the site of the calamity. And this, too, is a nice, upper-middle class neighbourhood.

But then something unexpected happens. 'It was like an instant reflex,' Ruben Abrahams, owner of a real estate agency on the corner, later tells a local TV reporter. 'Car in the water? That can't be good.'[16] He runs to get a hammer from his office toolbox and then sprints right into the icy canal.

A tall, athletic guy with greying stubble, Ruben meets me one cold January day to show me where it all happened. 'It was one of those bizarre coincidences,' he tells me, 'where everything came together in a split second.'

When Ruben jumps into the canal, Rienk Kentie – also a bystander – is already swimming towards the sinking automobile, and Reinier Bosch – yet another bystander – is in the water, too. At the last instant, a woman had handed Reinier a brick, something that moments later will prove crucial. Wietse Mol – bystander number four – grabs an emergency hammer from his car and is the last to dive in.

'We began bashing on the windows,' Ruben recounts.[17] Reinier tries to smash one of the side windows, but no luck. Meanwhile, the car tilts and dips, nose down. Reinier brings the brick crashing down hard on the back window. Finally, it cracks.

After that, everything happens very fast. 'The mother passed her child to me through the back window,' Ruben continues. For a moment, the kid gets stuck, but a few seconds later Ruben and Reinier manage to work the toddler free. Reinier swims the child to safety. With the mother still inside, the car is inches away from going under. Just in time, Ruben, Rienk and Wietse help her get out.

Not two seconds later, the car vanishes into the inky waters of the canal.

By that time, a whole crowd of bystanders has gathered along the waterside. They help lift the mother and child and four men out of the water and wrap them in towels.

The whole rescue operation was over in less than two minutes. In all that time, the four men – complete strangers to one another – never exchanged a word. If any of them had hesitated for even a split second longer, it would have been too late. If all four had not jumped in, the rescue may well have failed. And if that nameless bystander had not handed Reinier a brick at the last instant, he wouldn't have been able to smash the back window and get the mother and child out.

In other words, Sanne and her toddler survived not in spite of the large number of bystanders, but *because* of them.

4

Now, you could think – touching story, sure, but it's probably the exception to the bystander rule. Or maybe there's something special about the Dutch culture, or this neighbourhood in Amsterdam, or even these four men, that accounts for the anomaly?

On the contrary. Though the bystander effect may still be taught in many textbooks, a meta-analysis published in 2011 has shed new light on what bystanders do in emergencies. Meta-analysis is research about research, meaning it analyses a large group of other studies. This meta-analysis reviewed the 105 most important studies on the bystander effect from the past fifty years, including that first experiment by Latané and Darley (with students in a room).[18]

Two insights came out of this study-of-studies. One: the bystander effect exists. Sometimes we think we don't need to intervene in emergencies because it makes more sense to let somebody else take charge. Sometimes we're afraid to do the wrong thing and don't intervene for fear of censure. And sometimes we simply don't think there's anything wrong, because we see that nobody else is taking action.

And the second insight? If the emergency is life-threatening (somebody is drowning or being attacked) and if the bystanders can communicate with one another (they're not isolated in separate rooms), then there's an *inverse* bystander effect. 'Additional bystanders,' write the article's authors, 'even lead to more, rather than less, helping.'[19]

And that's not all. A few months after interviewing Ruben about his spontaneous rescue effort, I arrange to meet Danish psychologist Marie Lindegaard at a café in Amsterdam. Still shaking off raindrops, she sits down, opens her laptop, drops a stack of papers in front of me and launches into a lecture.

Lindegaard was one of the first researchers to ask why we think up all these convoluted experiments, questionnaires and interviews. Why don't we simply look at real footage of real people in real situations? After all, modern cities are chock-a-block with cameras.

Great idea, Marie's colleagues answered, but you'll never be able to get your hands on that footage. To which Marie replied: we'll see about that. These days, Marie has a database containing over a thousand videos from Copenhagen, Cape Town, London and Amsterdam. They record brawls, rapes and attempted murders, and her findings have started a minor revolution in the social sciences.

She pushes her laptop towards me. 'Look, tomorrow we're submitting this article to a leading psychology journal.'[20]

I read the working title: '*Almost Everything You Think You Know About the Bystander Effect is Wrong.*'

Lindegaard scrolls down and points to a table. 'And look, here you can see that in 90 per cent of cases, people help each other out.'

Ninety per cent.

5

It's no mystery, then, why Ruben, Reinier, Rienk and Wietse dived into the ice-cold waters of an Amsterdam canal that February afternoon. It was the natural response. The question now is: what happened on 13 March 1964, the night Kitty Genovese was murdered? How much of that well-known story is true?

One of the first people to question the apathy of the eyewitnesses was a newcomer to Kew Gardens, Joseph De May. The amateur historian moved there ten years after Kitty's death and was intrigued by the murder that had made the neighbourhood infamous. De May decided to do some research of his own. He started to go through the archives and turned up faded photographs and old newspapers and police reports. Piece by piece, as he began putting everything together, a picture emerged of what really happened.

Let's take it again from the top. Here are the events of 13 March 1964, this time relying on the painstaking investigation carried out by De May and others who followed in his wake.[21]

It's 3:19 a.m. when a horrifying scream breaks the silence on Austin Street. But it's cold outside, and most residents have their windows shut. The street is poorly lit. Most people who look outside don't notice anything odd. A few make out the silhouette of a woman lurching down the street and assume she must be drunk. That wouldn't be unusual, as there's a bar just up the street.

Nevertheless, at least two residents pick up the phone and call the police. One of them is the father of Michael Hoffmann, who will later join the force himself, and the other is Hattie Grund, who lives in an apartment nearby. 'They said,' she repeats years later, 'we already got the calls.'[22]

But the police don't come.

The police don't come? Why didn't they tear out of the station, sirens blaring?

Based on those first calls, the dispatcher may have assumed this was a marital spat. Hoffmann, now retired from the force, thinks that's why they were so slow to arrive on the scene. Bear in mind these were the days when people didn't pay much attention to a husband beating his wife, the days when spousal rape wasn't even a criminal offence.

But what about those thirty-eight eyewitnesses?

This notorious number, which would turn up in everything from songs and plays to blockbusters and bestsellers, comes from a list of all the people questioned in the case by police detectives. And the vast majority of the names on that list *were not eyewitnesses*. At most they'd heard something, but some hadn't woken up at all.

There were two clear exceptions. One was Joseph Fink, a neighbour in the building. Fink was an odd, solitary man who was known to hate Jews (the local kids called him 'Adolf'). He was wide awake when it happened, he saw the first attack on Kitty and he did nothing.

The other person who abandoned Kitty to her fate was Karl Ross, a neighbour who was friends with her and Mary Ann. Ross personally witnessed the second attack in the stairwell (in reality, there were two attacks, not three), but he panicked and left. Ross was also the man who told police he 'didn't want to get involved' – but what he meant was that he didn't want publicity. He was drunk that night, and he was afraid it would come out that he was gay.

Homosexuality was strictly illegal in those days, and Ross was terrified both of the police and of papers like the *New York Times*, which stigmatized homosexuality as a dangerous disease.[23] In 1964, gay men were still routinely brutalised by police, and the paper regularly portrayed homosexuality as a plague. (Abe Rosenthal in particular, the editor who made Kitty famous, was a notorious homophobe. Not long before Kitty's murder, he'd published another piece: GROWTH OF OVERT HOMOSEXUALITY IN CITY PROVOKES WIDE CONCERN.[24])

Of course, none of this excuses Karl Ross's negligence. Even if he was drunk and scared, he should have done more to help his friend. Instead, he phoned another friend, who immediately urged him to call the cops. But Ross didn't dare do so from his own apartment, so instead he climbed over the roof to his next-door neighbour's house and she woke the woman who lived next door to her.

That woman was Sophia Farrar. When Sophia heard that Kitty lay bleeding downstairs, she didn't hesitate for a second. She ran out of the apartment, leaving her husband still pulling on his trousers and calling after her to wait. For all Sophia knew, she could have been rushing straight into the arms of the murderer, but that didn't stop her. 'I ran to help. It seemed the natural thing to do.'[25]

When she opened the door to the stairwell where Kitty lay, the murderer was gone. Sophia put her arms around her friend, and Kitty relaxed for a moment, leaning into her. This, then, is how Catherine Susan Genovese really died: wrapped in her neighbour's embrace. 'It would have made such a difference to my family,' her brother Bill said when he heard this story many years later, 'knowing that Kitty died in the arms of her friend.'[26]

Why was Sophia forgotten?

Why wasn't she mentioned in any of the papers?

The truth is pretty disheartening. According to her son, 'My mom spoke to one woman from a newspaper back then', but when the article appeared the next day it said Sophia hadn't wanted to get involved. Sophia was furious when she read the piece and swore never to speak to a journalist again.

Sophia wasn't the only one. In fact, dozens of Kew Garden residents complained that their words kept getting twisted by the press, and many of them wound up moving out of the area. Journalists, meanwhile, kept dropping by. On 11 March 1965, two days before the first anniversary of Kitty's death, one reporter thought it would be a good joke to go to Kew Gardens and scream bloody murder in the middle of the night. Photographers stood with their cameras ready to capture residents' reactions.

The whole situation seems insane. In the same years that activism began brewing in New York City, that Martin Luther King was awarded the Nobel Peace Prize, that millions of Americans began marching in the streets and that Queens counted more than two hundred community organisations, the press developed an obsession with what it trumpeted as an 'epidemic of indifference'.

There was one journalist, a radio reporter named Danny Meenan, who was sceptical of the story about the disinterested bystanders. When he checked the facts, he found that most of the eyewitnesses thought they had seen a drunken woman that night. When Meenan asked the reporter at the *New York Times* why he hadn't put that information in his piece, his answer was, 'It would have ruined the story.'[27]

So why did Meenan keep this to himself? Self-preservation. In those days, no lone journalist would get it into his head to contradict the world's most powerful newspaper – not if they wanted to keep their job.

When another reporter sounded a critical note a few years later, he got a furious phone call from Abe Rosenthal at the *New York Times*. 'Do you realize that this story has become emblematic of a situation in America?' the editor screamed down the line. 'That it's become the subject of sociology courses, books, and articles?'[28]

It's shocking how little of the original story holds up. On that fateful night, it wasn't ordinary New Yorkers, but the authorities who failed. Kitty didn't die all alone, but in the arms of a friend. And when it comes down to it, the presence of bystanders has precisely the opposite effect of what science has long insisted. We're not alone in the big city, on the subway, on the crowded streets. We have each other.

And Kitty's story doesn't end there. There was one final, bizarre twist.

Five days after Kitty's death, Raoul Cleary, a Queens resident, noticed a stranger in his street. He was coming out of a neighbour's house in broad daylight, carrying a television set. When Raoul stopped him, the man claimed to be a mover.

But Raoul was suspicious and phoned a neighbour, Jack Brown.

'Are the Bannisters moving?' he asked.

'Absolutely not,' Brown answered.

The men didn't hesitate. While Jack disabled the man's vehicle, Raoul called the police, who arrived to arrest the burglar the moment he re-emerged. Just hours later, the man confessed. Not only to breaking and entering, but also to the murder of a young woman in Kew Gardens.[29]

That's right, Kitty's murderer was apprehended thanks to the intervention of two bystanders. Not a single paper reported it.

This is the real story of Kitty Genovese. It's a story that ought to be required reading not only for first-year psychology students, but also for aspiring journalists. That's because it teaches us three things. One, how out of whack our view of human nature often is. Two, how deftly journalists push those buttons to sell sensational stories. And, last but not least, how it's precisely in emergencies that we can count on one another.

As we look out across the water in Amsterdam, I ask Ruben Abrahams if he feels like a hero after his dip in the canal. 'Nah,' he shrugs, 'you've got to look out for each other in life.'

Part Three

WHY GOOD PEOPLE TURN BAD

'I have striven not to laugh at human actions, not to weep at them, nor to hate them, but to understand them.'

Baruch Spinoza (1632–77)

Not long ago I sat down with a book I wrote back in 2013 in my native Dutch, whose title translates as *The History of Progress*. Rereading it was an uncomfortable experience. In that book I dished up Philip Zimbardo's Stanford Prison 'research' without an ounce of criticism, as proof that good people can spontaneously turn into monsters. Clearly something about this observation had been irresistible to me.

I wasn't the only one. Since the end of the Second World War, countless variations on veneer theory have been put forth, supported by evidence that seemed increasingly iron-clad. Stanley Milgram demonstrated it using his shock machine. The media shouted it from the rooftops following Kitty Genovese's death. And William Golding and Philip Zimbardo rode the theory to worldwide fame. Evil was thought to simmer just beneath the surface in every human being, just as Thomas Hobbes had argued three hundred years earlier.

But now the archives of the murder case and the experiments have been opened up, and it turns out we had it back to front all along. The guards in Zimbardo's prison? They were actors playing parts. The volunteers at Milgram's shock machine? They wanted to do right. And Kitty? She died in the arms of a neighbour.

Most of these people, it seems, just wanted to help out. And if anyone failed, it was the people in charge – the scientists and the lead editors, the governors and the police chiefs. They were the Leviathans that lied and manipulated. Instead of shielding

subjects from their ostensibly wicked inclinations, these authorities did their best to pit people against one another.

This brings us back to the fundamental question: why do
people do evil things? How come *Homo puppy*, that friendly
biped, is the only species that's built jails and gas chambers?

In the previous chapters, we learned that humans may be
tempted by evil when it masquerades as good. But this finding
immediately raises another question: why has evil grown so
skilled at fooling us over the course of history? How did it
manage to get us to the point that we would declare war on one
another?

I keep thinking of an observation made by Brian Hare, our
puppy expert from Chapter 3, who said: 'The mechanism that
makes us the kindest species also makes us the cruelest species
on the planet.'

For most of human history, as we've seen, this statement
didn't apply. We haven't always been so cruel. For tens of
thousands of years, we roamed the world as nomads and kept
well clear of conflicts. We didn't wage war and we didn't build
concentration camps.

But what if Hare is on to something? What if his observation does apply to the last 5 per cent of human history, from
the time we began living in permanent settlements? It can be no
accident that the first archaeological evidence for war suddenly
appears approximately ten thousand years ago, coinciding with
the development of private property and farming. Could it be
that at this juncture we chose a way of life for which our bodies
and minds were not equipped?

Evolutionary psychologists refer to this as a *mismatch*,
meaning a lack of physical or mental preparation for modern
times. The most familiar illustration is obesity: where as hunter-
gatherers we were still slim and fit, these days more people

worldwide are overweight than go hungry. We regularly feast on sugars and fats and salts, taking in far more calories than our bodies need.

So why do we keep right on eating? Simple: our DNA thinks we're still running around in the jungle. In prehistory it made good sense to stuff yourself anytime you stumbled on a heavily laden fruit tree. That didn't happen very often, so building an extra layer of body fat was basically a self-preservation strategy.[1] But now, in a world awash with cheap, fast food, piling on extra fat is more like self-sabotage.

Is this how we should also be thinking about the darkest chapters of human history? Might they, too, be the result of a dramatic mismatch? And could that explain how modern-day *Homo puppy* came to be capable of the most heinous cruelty? In that case, there would have to be some aspect of our nature that misfires when confronted with life in the modern, 'civilised' world – some inclination that didn't bother us for millennia and then suddenly revealed its drawbacks.

Something, but what?

In the next three chapters, this is my quest. I'll introduce you to a young American who was determined to understand why the Germans fought so tirelessly right up to the very end of the Second World War (Chapter 10). We'll dive into psychological research on the cynicism that comes with power (Chapter 11). And then we'll take on the ultimate question: what kind of society can you get when people acknowledge the mismatch and choose to adopt a new, realistic view of humanity?

10

How Empathy Blinds

I

Morris Janowitz was twenty-two at the outbreak of the Second World War. A year later, a draft notice from the US Army arrived on his doormat. Finally. Morris was on fire to enlist. As the son of Jewish refugees from Poland, he couldn't wait to don a uniform and help beat the Nazis.[1]

The young man had long been fascinated by the social sciences. And now, having just graduated from college at the top of his class, he could put his expertise to work for the cause. Morris wasn't being sent into combat with a helmet and rifle, but wielding pen and paper. He was stationed at the Psychological Warfare Division in London.

At the agency's headquarters near Covent Garden, Morris joined dozens of top scientists, many of whom would later go on to illustrious careers in sociology and psychology. But this was not the time for abstract theorising. Science had been called to action. There was work to be done and not a moment to lose.

While the smartest physicists were cooking up the first atomic bomb in the town of Los Alamos in the American South West, and the cleverest mathematicians were cracking the Germans' Enigma Code in the English countryside at Bletchley Park, Morris and his colleagues were grappling with the toughest task of all.

They had to unravel the mystery of the Nazi mind.

By early 1944 there was one conundrum that had scientists stumped. Why did the Germans continue to fight so hard? Why weren't more of their soldiers laying down their arms and conceding defeat?

Anyone surveying the battlefield could see what the outcome would be. Vastly outnumbered, the Germans were sandwiched between the advancing Russians in the east and an imminent Allied invasion in the west. Did the average German on the ground not realise how heavily the odds were stacked against them, the Allies wondered? Had they been so thoroughly brainwashed? What else could explain why the Germans continued fighting to the last gasp?

From the outset of the war, most psychologists firmly believed that one factor outweighed every other in determining an army's fighting power. Ideology. Love of one's country, for example, or faith in one's chosen party. The soldiers who were most thoroughly convinced *they* stood on the right side of history and that *theirs* was the legitimate worldview would – so the thinking went – put up the best fight.

Most experts agreed that the Germans were in essence possessed. This explained their desertion rate that approached zero, and why they fought harder than the Americans and the British. So much harder, historians calculated after the war, that the average Wehrmacht soldier inflicted 50 per cent more casualties than his Allied counterpart.[2]

German soldiers were better at just about everything. Whether attacking or defending, with air support or without, it made no difference. 'The inescapable truth,' a British historian later observed, 'is that Hitler's Wehrmacht was the outstanding fighting force of World War II, one of the greatest in history.'[3]

And it was this army's morale that the Allies had to find a way to break. Morris and his team knew they needed to think big – very big. On the Psychological Warfare Division's recommendation, tens of millions of propaganda leaflets were dropped over enemy territory, reaching as much as 90 per cent of the German forces stationed in Normandy after D-Day. The message they rained down over and over was that the German position was hopeless, the Nazi philosophy despicable and the Allied cause justified.

Did it work? Morris Janowitz didn't have the foggiest idea. There was little chance of finding out from the confines of his desk, so he and fellow researcher Edward Shils decided to draw up a detailed survey to measure the leaflet campaign's effect. A few months later, Morris set off for liberated Paris to interview hundreds of German prisoners of war. It was during these talks that it started to dawn on him.

They'd got it all wrong.

For weeks Morris interviewed one German captive after another. He kept hearing the same responses. No, it wasn't the draw of Nazi ideology. No, they didn't have any illusions that they could still somehow win. No, they hadn't been brainwashed. The real reason why the German army was capable of putting forth an almost superhuman fight was much simpler.

Kameradschaft.

Friendship.

All those hundreds of bakers and butchers, teachers and tailors; all those German men who had resisted the Allied advance tooth and nail had taken up arms for one another. When it came down to it, they weren't fighting for a Thousand-Year Reich or for *Blut und Boden* – 'blood and soil' – but because they didn't want to let down their mates.

'Nazism begins ten miles behind the front line,' scoffed one German prisoner, whereas friendship was right there in every bunker and trench.[4] The military commanders were well aware of this, and, as later historians discovered, used it to their advantage.⁵ Nazi generals went to great lengths to keep comrades together, even withdrawing whole divisions for as long as it took new recruits to form friendships, and only then sent everyone back into the fray.

Envisaging the strength of this camaraderie in the Wehrmacht isn't easy. After all, we have been inundated for decades with Hollywood epics about Allied courage and German insanity. That our boys laid down their lives for one another? Logical. That they grew to be inseparable bands of brothers? Makes sense. But to imagine the same of the German hordes? Or, worse, that the Germans might have forged friendships that were even stronger? And that it was *because* of those friendships that their army was better?

Some truths are almost too painful to accept. How could it be that those monsters were also motivated by the good in humanity – that they, too, were fuelled by courage and loyalty, devotion and solidarity?

Yet that's precisely what Morris Janowitz concluded.

When the researchers at the Psychological Warfare Division put two and two together they suddenly understood why their propaganda campaign had virtually no impact. Writing about the effect of the millions of leaflets dropped behind enemy lines, Janowitz and Shils noted that 'Much effort was devoted to ideo-logical attacks on German leaders, but only about five per cent of the prisoners mentioned this topic [when questioned].'⁶

In fact, most Germans didn't even remember that the leaflets criticised National Socialism. When the researchers asked one German sergeant about his political views, the man burst out laughing: 'When you ask such a question, I realize well that you have no idea of what makes a soldier fight.'⁷

Tactics, training, ideology – all are crucial for an army, Morris and his colleagues confirmed. But ultimately, an army is only as strong as the ties of fellowship among its soldiers. Camaraderie is the weapon that wins wars.

These findings were published shortly after the war and would be reiterated by many subsequent studies. But the clincher came in 2001 when historians discovered 150,000 typed pages of conversations overheard by the US Secret Service. These were transcripts of things said by some four thousand Germans at a wire-tapped POW camp at Fort Hunt in Washington DC. Their talk opened an unprecedented window into the lives and minds of ordinary Wehrmacht servicemen.

The Germans, these transcripts showed, had a tremendous 'martial ethos' and placed a high value on qualities such as loyalty, camaraderie and self-sacrifice. Conversely, anti-Jewish sentiment and ideological purity played only a small role. 'As the wiretap transcripts from Fort Hunt show,' writes one German historian, 'ideology played at most a subordinate role in the consciousness of most Wehrmacht members.'[8]

The same was true of Americans fighting in the Second World War. In 1949, a team of sociologists published the results of a vast survey among some half million US war veterans, which revealed they had not been motivated primarily by idealism or ideology. An American soldier wasn't fuelled by patriotic spirit any more than a British one was by democratic rule of law. It wasn't so much for their countries that these men fought as for their comrades.[9]

So deep were these ties that they could lead to some peculiar situations. Servicemen would turn down promotion if it meant transferring to a different division. Many who were injured and sick refused leave because they didn't want a new recruit to take their place. And there were even men who sneaked out of their infirmary beds to escape back to the front.

'Time and again,' one sociologist noted in surprise, 'we encountered instances when a man failed to act in accordance with his own self-interests [for fear of] letting the other guys down.'[10]

2

It took me a long time to get to grips with this idea.

As a teenager growing up in Holland, I'd pictured the Second World War as a kind of twentieth-century *Lord of the Rings* – a thrilling battle between valiant heroes and evil villains. But Morris Janowitz showed that something altogether different was going on. The origins of evil, he discovered, lay not in the sadistic tendencies of degenerate bad guys, but in the solidarity of brave warriors. The Second World War had been a heroic struggle in which friendship, loyalty, solidarity – humanity's best qualities – inspired millions of ordinary men to perpetrate the worst massacre in history.

Psychologist Roy Baumeister calls the fallacious assumption that our enemies are malicious sadists 'the myth of pure evil'. In reality, our enemies are just like us.

This applies even to terrorists.

They're also like us, experts emphasise. Of course, it's tempting to think that suicide bombers must be monsters. Psychologically, physiologically, neurologically – they must be every kind of screwed up. They must be psychopaths, or maybe they never went to school, or grew up in abject poverty – there must be something to explain why they deviate so far from the average person.

Not so, say sociologists. These stoic data scientists have filled miles of Excel sheets with the personality traits of people who have blown themselves up, only to conclude that, empirically, there is no such thing as an 'average terrorist'. Terrorists span the spectrum from highly to hardly educated, from rich to poor, silly to serious, religious to godless. Few have mental illnesses and traumatic childhoods also appear to be rare. After an act of terror the media often show the shocked response of neighbours, acquaintances and friends, who, when asked about the suicide bomber, remember them as 'friendly' or 'a nice guy'.[11]

If there is any one characteristic that terrorists share, say experts, it's that they're so easily swayed. Swayed by the opinions of other people. Swayed by authority. They yearn to be seen and want to do right by their families and friends.[12] 'Terrorists don't kill and die just for a cause,' one American anthropologist notes. 'They kill and die for each other.'[13]

By extension, terrorists also don't radicalise on their own, but alongside friends and lovers. A large share of terrorist cells are quite literally 'bands of brothers': no fewer than four pairs of brothers were involved in the 2001 attacks on the Twin Towers, the 2013 Boston Marathon bombers were brothers, and so were Salah and Brahim Abdeslam, responsible for the Bataclan slaughter in Paris in 2015.[14]

It's no mystery why terrorists act together: brutal violence is frightening. As much as politicians talk about 'cowardly acts', in truth it takes a lot of nerve and determination to fight to the death. 'It's easier,' one Spanish terrorism expert points out, 'to take that leap accompanied by someone you trust and love.'[15]

When terrorists strike, the news media primarily focuses on the sick ideology that supposedly fuelled the attack. And, of course, ideology does matter. It mattered in Nazi Germany, and it certainly matters for the leaders of terrorist organisations like Al-Qaeda and Islamic State (IS), many of whom have been moulded by a youth spent devouring books on radical Islam (such as Osama bin Laden, a known bookworm).[16]

But research shows that for the foot soldiers of these organisations, ideology plays a remarkably small role. Take the thousands of Jihadists who set out for Syria in 2013 and 2014. Three-quarters were recruited by acquaintances and friends. Most, according to responses to a leaked IS poll, scarcely knew the first thing about the Islamic faith.[17] A few wisely bought *The Koran for Dummies* just before their departure. For them, says a CIA officer, 'Religion is an afterthought.'[18]

The thing we need to understand is that most of these terrorist agents were not religious fanatics. They were the best of friends. Together, they felt a part of something bigger, that their lives finally held meaning. At last they were the authors of their own epic tale.

And no, this is in no way an excuse for their crimes. It's an explanation.

3

In the autumn of 1990, a new research centre opened at the university where Stanley Milgram conducted his shock experiments thirty years earlier. Yale's Infant Cognition Center – or 'Baby Lab' as it's known – is doing some of the most exciting research around. The questions investigated here trace their roots right back to Hobbes and Rousseau. What is human nature? What's the role of nurture? Are people fundamentally good or bad?

In 2007, Baby Lab researcher Kiley Hamlin published the results of a groundbreaking study. She and her team were able to demonstrate that infants possess an innate sense of morality. Infants as young as six months old can not only distinguish right from wrong, but they also prefer the good over the bad.[19]

Perhaps you're wondering how Hamlin could be so sure. Babies can't do much on their own, after all. Mice can run mazes, but babies? Well, there is one thing they can do: babies can watch. So the researchers put on a puppet show for their pint-sized subjects (six and ten months old), featuring one puppet that acted helpful and another that behaved like a jerk. Which puppet would the infants then reach for?

You guessed it: infants favoured the helper puppet. 'This wasn't a subtle statistical trend,' one of the researchers later wrote. 'Just about all of the babies reached for the good guy.'[20] After centuries of speculating how babies see the world, here was cautious evidence to suggest we possess an innate moral compass and *Homo puppy* is not a blank slate. We're born with a preference for good; it's in our nature.

And yet as I dug deeper into the world of baby research, I soon began to feel less optimistic.

The thing is, human nature has another dimension. A few years after this first experiment, Hamlin and her team came up with a variation.[21] This time, they offered infants a choice between Graham Crackers and green beans to establish which they preferred. Then they presented them with two puppets: one that liked the crackers, the other the beans. Once again, they observed which puppet the babies favoured.

Not surprisingly, the overwhelming majority gravitated towards the puppet that shared their own taste. More surprising was that this preference persisted even after the like-minded puppet was revealed to be mean-spirited and the other puppet nice. 'What we find over and over again,' said one of Hamlin's colleagues, 'is that babies will choose the individual who is actually mean [but similar to them] to the [nice] one who had the different opinion to themselves.'[22]

How depressing can you get?

Even before we learn to speak, we seem to have an aversion to the unfamiliar. Researchers at the Baby Lab have done dozens of experiments which furthermore show that babies don't like unfamiliar faces, unknown smells, foreign languages, or strange accents. It's as though we're all born xenophobes.[23]

Then I began to wonder: could this be a symptom of our fatal mismatch? Could our instinctive preference for what we know have been no big deal for most of human existence, only to become a problem with the rise of civilisation? For more than 95 per cent of our history, after all, we were nomadic foragers. Any time we crossed paths with a stranger we could stop to chat and that person was a stranger no more.

Nowadays, things are very different. We live in anonymous cities, some of us among *millions* of strangers. Most of what we know about other people comes from the media and from journalists, who tend to zoom in on the bad apples. Is it any wonder we've become so suspicious of strangers? Could our innate aversion to the unfamiliar be a ticking time bomb?

Since that first study by Kiley Hamlin, many more have been conducted to test babies' sense of morality. It's a fascinating field of research, albeit one that's still in its, um, infancy. The big stumbling block with this kind of research is that babies are easily distracted, which makes it difficult to design reliable experiments.[24]

Fortunately, by the time we reach eighteen months, humans are a good deal smarter and therefore easier to study. Take the work of the German psychologist Felix Warneken. As a Ph.D. student, he became interested in investigating how helpful toddlers were. His supervisors rejected the whole idea, believing – as was common in the early 2000s – that toddlers were basically walking egos. But Warneken was not to be deterred and set up a series of experiments that would eventually be replicated around the world.[25]

Across the board, their results were the same. The experiments revealed that even at the tender age of eighteen months children are only too eager to help others, happily taking a break from fun and games to lend a hand, helping a stranger even when you throw a ball pit into the mix.[26] And they want nothing in return.[27]

But now for some bad news. After learning about Felix Warneken's uplifting research, I also encountered quite a few studies whose findings are less rosy, showing that children can be turned against each other. We saw this with Muzafer Sherif's Robbers Cave Experiment (see Chapter 7), and it was demonstrated again by a notorious experiment from the 1960s, launched the day after the assassination of Martin Luther King Jr.

On 5 April 1968, Jane Elliott decided to give her class of third-graders at a small school in Riceville, Iowa, a hands-on lesson in racism.

'The brown-eyed people are the better people in this room,' Elliott began. 'They are cleaner and they are smarter.' In capital letters she wrote the word MELANIN on the chalkboard, explaining that this is the chemical that makes people smart. As kids with brown eyes had more of it, they were also more intelligent, whereas their blue-eyed counterparts 'sit around and do nothing'.[28]

It didn't take long for the Brownies to start talking down to the Blueys, and then for the Blueys to lose their confidence. A normally smart blue-eyed girl began making mistakes during a maths lesson. During the break afterwards, she was approached by three brown-eyed friends. 'You better apologize to us for getting in our way,' one of them said, 'because we're better than you are.'[29]

When Elliott appeared as a guest on the popular *Tonight Show Starring Johnny Carson* a few weeks later, white America was outraged. 'How dare you try this cruel experiment out on white children,' wrote one angry viewer. 'Black children grow up accustomed to such behavior, but white children, there's no way they could possibly understand it. It's cruel to white children and will cause them great psychological damage.'[30]

Jane Elliott continued to fight this kind of racism all her life. But it's crucial to bear in mind that hers was not a scientific set-up. She took great pains to pit her pupils against one another, for example forcing the blue-eyed kids to sit at the back of the classroom, giving them less breaktime and not allowing them to play with their brown-eyed peers. Her experiment didn't answer the question of what happens when you split kids into groups, but don't intervene in any other way.

In the autumn of 2003, a team of psychologists designed a study to do precisely that. They asked two day-care centres in Texas to dress all their children, aged three to five, in different coloured shirts, either red or blue. After only three weeks, the researchers were already able to draw some conclusions.[31] To begin with, as long as the adults ignored the difference in colours, the toddlers didn't pay them any attention either. Nonetheless, the children *did* develop a sense of group identity. In conversations with the researchers, they called their own colour 'smarter' and 'better'. And in a variation on the experiment where adults underscored the differences ('Good morning, reds and blues!'), this effect was even stronger.

In a subsequent study, a group of five-year-olds was similarly dressed in red or blue shirts and then shown photographs of peers who were wearing either the same or the other colour. Even without knowing anything else about the pictured individuals, the study subjects had a considerably more negative view of the children shown wearing a different colour to their own. Their perceptions, observed the researchers, were 'pervasively distorted by mere membership in a social group, a finding with disturbing implications'.[32]

The harsh lesson is that toddlers are not colour-blind. Quite the reverse: they're more sensitive to differences than most adults realise. Even when people try to treat everyone as equals and act as though variations in skin colour, appearance, or wealth don't exist, children still perceive the difference. It seems we're born with a button for tribalism in our brains. All that's needed is for something to switch it on.

4

As I read about the split nature of infants and toddlers – basically friendly, but with xenophobic tendencies – I was reminded of the 'love hormone' oxytocin. That's the stuff found in high concentrations in Lyudmila Trut's foxes in Siberia (see Chapter 3). Scientists now know that this hormone, which plays a crucial role in love and affection, can also make us distrustful of strangers.

Could oxytocin help explain why good people do bad things? Do strong ties to our own group predispose us to feel animosity towards others? And could the sociability that enabled *Homo puppy* to conquer the world also be the source of humankind's worst transgressions?[33]

Initially this line of thinking struck me as rather unlikely. After all, people have another impressive instinct rooted deep in our puppyish nature: the ability to feel empathy. We can step out of our bubble and into someone else's shoes. We're hardwired to feel, at an emotional level, what it's like to be the stranger.

Not only *can* we do this, but we're good at it. People are emotional vacuum cleaners, always sucking up other people's feelings. Just think how easily books and movies can make us laugh or cry. For me, sad movies on flights are always the worst

(I'm constantly pressing pause so fellow passengers won't feel the need to comfort me).[34]

For a long time I thought this fabulous instinct for feeling another person's pain could help bring people closer together. What the world needed, surely, was a lot more empathy. But then I read a new book by one of those baby researchers.

When people ask Professor Bloom what his book's about, he'll say:

'It's about empathy.'

They smile and nod – until he adds:

'I'm against it.'[35]

Paul Bloom isn't joking. According to this psychologist, empathy isn't a beneficent sun illuminating the world. It's a spotlight. A searchlight. It singles out a specific person or group of people in your life, and while you're busy sucking up all the emotions bathed in that one ray of light, the rest of the world fades away.

Take the following study carried out by another psychologist. In this experiment, a series of volunteers first heard the sad story of Sheri Summers, a ten-year-old suffering from a fatal disease. She's on the waiting list for a life-saving treatment, but time's running out. Subjects were told they could move Sheri up the waiting list, but they're asked to be objective in their decision.

Most people didn't consider giving Sheri an advantage. They understood full well that every child on that list was sick and in need of treatment.

Then came the twist. A second group of subjects was given the same scenario, but was then asked to imagine how Sheri must be feeling: Wasn't it heartbreaking that this little girl was so ill? Turns out this single shot of empathy changed everything. The majority now wanted to let Sheri jump the line. If you think about it, that's a pretty shaky moral choice. The spotlight on

Sheri could effectively mean the death of other children who'd been on the list longer.[36]

Now you may think: 'Exactly! That's why we need *more* empathy.' We ought to put ourselves not only in Sheri's shoes, but in those of the other children on waiting lists all over the world. More emotions, more feelings, more empathy!

But that's not how spotlights work. Go ahead and try it: imagine yourself in the shoes of one other person. Now imagine yourself in the shoes of a hundred other people. And a million. How about seven billion?

We simply can't do it.

In practical terms, says Professor Bloom, empathy is a hopelessly limited skill.

It's something we feel for people who are close to us; for people we can smell, see, hear and touch. For family and friends, for fans of our favourite band, and maybe for the homeless guy on our own street corner. For cute puppies we can cuddle and pet, even as we eat animals mistreated on factory farms out of sight. And for people we see on TV – mostly those the camera zooms in on, while sad music swells in the background.

As I read Bloom's book, I began to realise that empathy resembles nothing so much as that modern-day phenomenon: the news. In Chapter 1, we saw that the news also functions like a spotlight. Just as empathy misleads us by zooming in on the specific, the news deceives us by zooming in on the exceptional.

One thing is certain: a better world doesn't start with more empathy. If anything, empathy makes us less forgiving, because the more we identify with victims, the more we generalise about our enemies.[37] The bright spotlight we shine on our chosen few makes us blind to the perspective of our adversaries, because everybody else falls outside our view.[38]

This is the mechanism that puppy expert Brian Hare talked about – the mechanism that makes us both the friendliest and the cruellest species on the planet. The sad truth is that empathy and xenophobia go hand in hand. They're two sides of the same coin.

5

So why do good people turn bad?

I think we can now start to frame an answer. In the Second World War, the soldiers of the Wehrmacht fought first and foremost for each other. Most were motivated not by sadism or a thirst for blood, but by comradeship.

Once in combat, we've seen that soldiers still find it hard to kill. In Chapter 4 we were in the Pacific with Colonel Marshall, who realised that the majority of soldiers never fired their guns. During the Spanish Civil War, George Orwell noticed the same thing, when one day he found himself overpowered by empathy:

> At this moment a man [...] jumped out of the trench and ran along the top of the parapet in full view. He was half-dressed and was holding up his trousers with both hands as he ran. I refrained from shooting at him. [...] I did not shoot partly because of that detail about the trousers. I had come here to shoot at 'Fascists'; but a man who is holding up his trousers isn't a 'Fascist', he is visibly a fellow creature, similar to yourself, and you don't feel like shooting at him.[39]

Marshall's and Orwell's observations illustrate the difficulty we have inflicting harm on people who come too close. There's something that holds us back, making us incapable of pulling the trigger.

There's one thing even harder to do than shoot, military historians have discovered: stabbing a fellow human being. Less than 1 per cent of injuries during the battles at Waterloo (1815) and the Somme (1916), for instance, were caused by soldiers wielding bayonets.[40] So all those thousands of bayonets displayed in hundreds of museums? Most were never used. As one historian notes, 'one side or the other usually recalls an urgent appointment elsewhere before bayonets cross'.[41]

Here, too, we've been misled by the television and movie industries. Series like *Game of Thrones* and movies like *Star Wars* would have us believe that skewering another person is a piece of cake. But in reality it's psychologically very hard to run through the body of another person.

So how do we account for the hundreds of millions of war casualties over the past ten thousand years? How did all those people die? Answering this question requires forensic examination of the victims, so let's take the causes of death of British soldiers in the Second World War as an example:[42]

Other: 1%
Chemical: 2%
Blast, crush: 2%
Landmine, booby trap: 10%
Bullet, anti-tank mine: 10%
Mortar, grenade, aerial bomb, shell: 75%

Notice anything? If there's one thing that ties these victims together, it's that most were eliminated remotely. The overwhelming majority of soldiers were killed by someone who pushed a button, dropped a bomb, or planted a mine. By someone who never saw them, certainly not while they were half-naked and trying to hold up their trousers.

Most of the time, wartime killing is something you do from far away. You could even describe the whole evolution of military technology as a process in which enemy lines have grown farther apart. From clubs and daggers to bows and arrows, and from muskets and cannon to bombs and grenades. Over the course of history, weaponry has got ever better at overcoming the central problem of all warfare: our fundamental aversion to violence. It's practically impossible for us to kill someone while looking them in the eyes. Just as most of us would instantly go vegetarian if forced to butcher a cow, most soldiers become conscientious objectors when the enemy gets too close.

Down the ages, the way to win most wars has been to shoot as many people as possible from a distance.[43] That's how the English defeated the French at Crécy and Agincourt during the Hundred Years' War (1337–1453), how the conquistadors conquered the Americas in the fifteenth and sixteenth centuries, and what the US military does today, with its legions of armed drones.

Aside from long-range weapons, armies also pursue means to increase *psychological* distance to the enemy. If you can dehumanise the other – say, by portraying them as vermin – it makes it easier to treat the other as if they are indeed inhuman.

You can also drug your soldiers to dull their natural empathy and antipathy towards violence. From Troy to Waterloo, from Korea to Vietnam, few armies have fought without the aid of intoxicants, and scholars now even think Paris might not have fallen in 1940 had the German army not been stoked on thirty-five million methamphetamine pills (aka crystal meth, a drug that can cause extreme aggression).[44]

Armies can also 'condition' their troops. The US Army started doing this after the Second World War on the recommendation of none other than Colonel Marshall. Vietnam recruits were immersed in boot camps that exalted not only a sense of brotherhood, but also the most brutal violence, forcing the men to scream 'KILL! KILL! KILL!' until they were hoarse. Second

World War veterans (most of whom had never learned to kill) were shocked when shown images of this brand of training.[45]

These days, soldiers no longer practise on ordinary paper bullseyes, but are drilled to fire instinctively at realistic human figures. Shooting a firearm becomes an automated, Pavlovian reaction you can perform without thinking. For snipers, the training's even more radical. One tried-and-tested method is to present a series of progressively more horrific videos while the trainee sits strapped to a chair and a special device ensures their eyes stay wide open.[46]

And so we're finding ways to root out our innate and deep-seated aversion to violence. In modern armies, comradeship has become less important. Instead we have, to quote one American veteran, 'manufactured contempt'.[47]

This conditioning works. Set soldiers trained using these techniques opposite an old-school army and the latter is crushed every time. Take the Falklands War (1982): though bigger in sheer numbers, the Argentine army with its old-fashioned training never had a chance against Britain's conditioned shooting machines.[48]

The American military also managed to boost its 'firing ratio', increasing the number of soldiers who shoot to 55 per cent in the Korean War and 95 per cent in Vietnam. But this came at a price. If you brainwash millions of young soldiers in training, it should come as no surprise when they return with post-traumatic stress disorder (PTSD), as so many did after Vietnam.[49] Innumerable soldiers had not only killed other people – something inside them had died, too.

Finally, there's one group which can easily keep the enemy at a distance: the leaders.

The commanders of armies and of terrorist organisations who hand down orders from on high don't have to stifle feelings of empathy for their opponent. And what's fascinating is that, while soldiers tend to be ordinary people, their leaders are a different story. Terrorism experts and historians consistently point out that people in positions of power have distinct psychological profiles. War criminals like Adolf Hitler and Joseph Goebbels are classic examples of power-hungry, paranoid narcissists.[50] Al-Qaeda and IS leaders have been similarly manipulative and egocentric, rarely troubled by feelings of compassion or doubt.[51]

This brings us to the next mystery. If *Homo puppy* is such an innately friendly creature, why do egomaniacs and opportunists, narcissists and sociopaths keep coming out on top? How can it be that we humans – one of the only species to blush – somehow allow ourselves to be ruled by specimens who are utterly shameless?

11

How Power Corrupts

I

If you want to write about power, there's one name you can't escape. He made a brief appearance in Chapter 3, where I discussed the theory that anyone who wants to achieve anything is best served weaving a web of lies and deception.

The name is Machiavelli.

In the winter of 1513, after yet another long night at the pub, a down-and-out city clerk started writing a pamphlet he called *The Prince*. This 'little whimsy of mine', as Machiavelli described it, was to become one of the most influential works in western history.[1] *The Prince* would wind up on the bedside tables of Emperor Charles V, King Louis XIV and General Secretary Stalin. The German Chancellor Otto van Bismarck had a copy, as did Churchill, Mussolini and Hitler. It was even found in Napoleon's carriage just after his defeat at Waterloo.

The big advantage of Niccolò Machiavelli's philosophy is that it's doable. If you want power, he wrote, you have to grab it. You must be shameless, unfettered by principles or morals. The ends justify the means. And if you don't look out for yourself, people will waltz right over you. According to Machiavelli, 'it can be said about men in general that they are ungrateful, fickle, dissembling, hypocritical, cowardly, and greedy'.[2] If someone does you a good turn, don't be fooled: it's a sham, for 'men never do anything good except out of necessity'.[3]

Machiavelli's book is often called 'realistic'. If you'd care to read it, just visit your nearest bookshop and head for the everseller. Or you can opt for one of the multitude of self-help books devoted to his philosophy, from *Machiavelli for Managers* to *Machiavelli for Moms*, or watch any number of plays, movies and TV series inspired by his ideas. *The Godfather*, *House of Cards*, *Game of Thrones* — all are basically footnotes to the work of this sixteenth-century Italian.

Given his theory's popularity, it makes sense to ask if Machiavelli was right. Must people shamelessly lie and deceive to gain and retain power? What does the latest science have to say?

Professor Dacher Keltner is the leading expert on applied Machiavellianism. When he first became interested in the psychology of power back in the nineties, he noticed two things. One: almost everybody believed Machiavelli was right. Two: almost nobody had done the science that could back it up.

Keltner decided to be the first. In what he termed his 'natural state' experiments, the American psychologist infiltrated a succession of settings where humans freely vie for dominion, from dorm rooms to summer camps. It was in precisely these kinds of places, where people meet for the first time, that he expected to see Machiavelli's timeless wisdom on full display.

He was disappointed. Behave as *The Prince* prescribes, Keltner discovered, and you'll be run right out of camp. Much as in prehistoric times, these mini-societies don't put up with arrogance. People assume you're a jerk and shut you out. The individuals who rise to positions of power, Keltner found, are the friendliest and the most empathic.[4] It's survival of the friendliest.

Now you may be thinking: this professor guy should swing by the office and meet my boss — that'll cure him of his little theory about nice leaders.

But hold on, there's more to this story. Keltner also studied the effects of power *once people have it*. This time he arrived at an altogether different conclusion. Perhaps most entertaining is his Cookie Monster study, named for the furry blue muppet from *Sesame Street*.[5] In 1998, Keltner and his team had small groups of three volunteers come into their lab. One was randomly assigned to be the group leader and they were all given a dull task to complete. Presently, an assistant brought in a plate containing five cookies for the group to share. All groups left one cookie on the plate (a golden rule of etiquette), but in almost every case the fourth cookie was scarfed down by the leader. What's more, one of Keltner's doctoral students noticed that the leaders also seemed to be messier eaters. Replaying the videos, it became clear that these 'cookie monsters' more often ate with their mouths open, ate more noisily and sprayed more crumbs on their shirts.

Maybe this sounds like your boss?

At first I was inclined to laugh off this kind of goofy experiment, but dozens of similar studies have been published in recent years from all over the world.[6] Keltner and his team did another one looking at the psychological effect of an expensive car. Here, the first set of subjects were put behind the wheel of a beat-up Mitsubishi or Ford Pinto and sent in the direction of a crosswalk where a pedestrian was just stepping off the curb. All the drivers stopped as the law required.

But then in part two of the study, subjects got to drive a snazzy Mercedes. This time, 45 per cent failed to stop for the pedestrian. In fact, the more expensive the car, the ruder the road manners.[7] 'BMW drivers were the worst,' one of the other researchers told the *New York Times*.[8] (This study has now been replicated twice with similar results.)[9]

Observing how the drivers behaved, Keltner eventually realised what it reminded him of. The medical term is 'acquired sociopathy': a non-hereditary antisocial personality disorder, first diagnosed by psychologists in the nineteenth century. It arises after a blow to the head that damages key regions of the brain and can turn the nicest people into the worst kind of Machiavellian.

It transpires that people in power display the same tendencies.[10] They literally act like someone with brain damage. Not only are they more impulsive, self-centred, reckless, arrogant and rude than average, they are more likely to cheat on their spouses, are less attentive to other people and less interested in others' perspectives. They're also more shameless, often failing to manifest that one facial phenomenon that makes human beings unique among primates.

They don't blush.

Power appears to work like an anaesthetic that makes you insensate to other people. In a 2014 study, three American neurologists used a 'transcranial magnetic stimulation machine' to test the cognitive functioning of powerful and less powerful people. They discovered that a sense of power disrupts what is known as *mirroring*, a mental process which plays a key role in empathy.[11] Ordinarily, we mirror all the time. Someone else laughs, you laugh, too; someone yawns, so do you. But powerful individuals mirror much less. It is almost as if they no longer feel connected to their fellow human beings. As if they've come unplugged.[12]

If powerful people feel less 'connected' to others, is it any wonder they also tend to be more cynical? One of the effects of power, myriad studies show, is that it makes you see others in a negative light.[13] If you're powerful you're more likely to think most people are lazy and unreliable. That they need to be supervised and monitored, managed and regulated, censored and told what to do. And because power makes you feel superior to other people, you'll believe all this monitoring should be entrusted to you.

Tragically, not having power has exactly the opposite effect. Psychological research shows that people who feel powerless also feel far less confident. They're hesitant to voice an opinion. In groups, they make themselves seem smaller, and they underestimate their own intelligence.[14]

Such feelings of uncertainty are convenient for those *in* power, as self-doubt makes people unlikely to strike back. Censorship becomes unnecessary, because people who lack confidence silence themselves. Here we see a nocebo in action: treat people as if they are stupid and they'll start to feel stupid, leading rulers to reason that the masses are too dim to think for themselves and hence they – with their vision and insight – should take charge.

But isn't it precisely the other way around? Isn't it power that makes us short-sighted? Once you arrive at the top, there's less of an impetus to see things from other perspectives. There's no imperative for empathy, because anyone you find irrational or irritating can simply be ignored, sanctioned, locked up, or worse. Powerful people don't have to justify their actions and therefore can afford a blinkered view.

That might also help explain why women tend to score higher than men on empathy tests. A large study at Cambridge University in 2018 found no genetic basis for this divergence, and instead attributed it to what scientists call *socialisation*.[15] Due to the way power has traditionally been distributed, it's mostly been up to women to understand men. Those persistent ideas about a superior female intuition are probably rooted in the same imbalance – that women are expected to see things from a male perspective, and rarely the other way around.

2

The more I found out about the psychology of power, the more I understood that power is like a drug – one with a whole catalogue of side effects. 'Power tends to corrupt, and absolute power corrupts absolutely,' British historian Lord Acton famously remarked back in the nineteenth century. There are few statements on which psychologists, sociologists and historians so unanimously agree.[16]

Dacher Keltner calls this 'the power paradox'. Scores of studies show that we pick the most modest and kind-hearted individuals to lead us. But once they arrive at the top, the power often goes straight to their heads – and good luck unseating them after that.

We need only look at our gorilla and chimpanzee cousins to see how tricky toppling a leader can be. In gorilla troops there's a single silverback dictator who makes all the decisions and has exclusive access to a harem of females. Chimp leaders also go to great lengths to stay on top, a position reserved for the male who's the strongest and most adept at forging coalitions.

'Entire passages of Machiavelli seem to be directly applicable to chimpanzee behavior,' biologist Frans de Waal noted in his book *Chimpanzee Politics*, published in the early eighties.[17] The alpha male – the prince – struts around like a he-man and manipulates the others into doing his bidding. His deputies help him hold the reins but could just as easily conspire to stab him in the back.

Scientists have known for decades that we share 99 per cent of our DNA with chimpanzees. In 1995, this inspired Newt Gingrich, then Speaker of the House of Representatives, to hand out dozens of copies of De Waal's book to his colleagues. The US Congress was to his mind not much different from a chimpanzee colony. At best, its members exercised a little more effort to hide their instincts.

What was not yet widely known at the time is that humans have another close primate relative that shares 99 per cent of our DNA. The bonobo. The first time Frans de Waal saw one was back in the early seventies, when they were still known as 'pygmy chimpanzees'. For a long time, chimps and bonobos were even thought to be the same species.[18]

In reality, bonobos are an altogether different creature. In Chapter 4 we saw that these apes have domesticated themselves, just like *Homo puppy*. The female of the species seem to have been key to this process, because, while not as strong as the males, they close ranks any time one of their own gets harassed by the opposite sex. If necessary, they bite his penis in half.[19] Thanks to this balance of power, bonobo females can pick and choose their own mates, and the nicest guys usually finish first.

(If you think all this emancipation makes for a dull sex life, think again: 'Bonobos behave as if they have read the *Kama Sutra*,' writes De Waal, 'performing every position and variation one can imagine.'[20] When two groups of bonobos first meet, it often ends in an orgy.)

Before we get too enthusiastic: humans are clearly not bonobos. Still, a growing body of research suggests that we have a lot more in common with these sociable apes than we do with Machiavellian chimpanzees. For starters, throughout most of human history our political systems much more closely resembled that of bonobos. Just recall the tactics of the !Kung tribe members (see Chapter 5): 'We refuse one who boasts, for someday his pride will make him kill somebody. So we always speak of his meat as worthless. This way we cool his heart and make him gentle.'

In an analysis of forty-eight studies on hunter-gatherer societies, an American anthropologist determined that Machiavellianism has almost always been a recipe for disaster. To illustrate this, here are some traits that, according to this scientist, were needed to get you elected leader in prehistory. You had to be:

Generous
Brave
Wise
Charismatic
Fair
Impartial
Reliable
Tactful
Strong
Humble [21]

Leadership was temporary among hunter-gatherers and decisions were made as a group. Anyone foolish enough to act as Machiavelli later prescribed was risking their life. The selfish and the greedy would get booted out of the tribe and faced likely starvation. After all, nobody wanted to share food with those who were full of themselves.

A further indication that human behaviour more closely resembles that of bonobos than chimpanzees is our innate aversion to inequality. Do a search for 'inequality aversion' in Google Scholar and you'll find more than ten thousand scientific articles about this primordial instinct. Children as young as three already divide a cake out equally, and at six would rather throw a slice away than let one person have a larger portion. [22] Like bonobos, humans share both fanatically and frequently.

That said, we also shouldn't exaggerate such findings. *Homo puppy* is not a natural-born communist. We're fine with a little inequality, psychologists emphasise, if we think it's justified. As long as things *seem* fair. If you can convince the masses that you're smarter or better or holier, then it makes sense that you're in charge and you won't have to fear opposition.

With the advent of the first settlements and growth in inequality, chieftains and kings had to start legitimising why they enjoyed more privileges than their subjects. In other words, they began engaging in propaganda. Where the chiefs of nomadic tribes were all modesty, now leaders began putting on airs. Kings proclaimed they ruled by divine right or that they themselves were gods.

Of course, the propaganda of power is more subtle these days, but that's not to say we no longer design ingenious ideologies to justify why some individuals 'deserve' more authority, status, or wealth than others. We do. In capitalist societies, we tend to use arguments of *merit*. But how does society decide who has the most merit? How do you determine who contributes most to society? Bankers or bin men? Nurses or the so-called disruptors who're always thinking outside the box? The better the story you spin about yourself, the bigger your piece of the pie. In fact, you could look at the entire evolution of civilisation as a history of rulers who continually devised new justifications for their privileges.[23]

But something strange is going on here. Why do we believe the stories our leaders tell us?

Some historians say it's because we're naive – and that might just be our superpower as a species.[24] Simply put, the theory goes like this: if you want to get thousands of strangers to work as a team, you need something to hold things together. This glue has to be stronger than friendliness, because although *Homo puppy*'s social network is the biggest of all primates, it isn't nearly large enough to forge cities or states.

Typically, our social circles number no more than about one hundred and fifty people. Scientists arrived at this limit in the 1990s, when two American researchers asked a group of volunteers to list all the friends and family to whom they sent Christmas cards. The average was sixty-eight households, comprising some one hundred and fifty individuals.[25]

When you start looking, this number turns up everywhere. From Roman legions to devout colonists and from corporate divisions to our real friends on Facebook, this magic threshold pops up all over the place and suggests the human brain is not equipped to juggle more than a hundred and fifty meaningful relationships.

The problem is that while a hundred and fifty guests make for a great party, it's nowhere near enough to build a pyramid or send a rocket to the moon. Projects on that scale call for cooperation in much larger groups, so leaders needed to incentivise us.

How? With myths. We learned to *imagine* kinship with people we'd never met. Religions, states, companies, nations – all of them really only exist in our minds, in the narratives our leaders and we ourselves tell. No one has ever met 'France' or shaken hands with 'the Roman Catholic Church'. But that doesn't matter if we sign on for the fiction.

The most obvious example of such a myth is, of course, God. Or call it the original Big Brother. Even as a teenager, I wondered why the Christian Creator I grew up with cared so much about humans and our mundane doings. Back then, I didn't know that our nomadic ancestors had a very different conception of the divine and that their gods took scant interest in human lives (see Chapter 5).

The question is: where did we get this belief in an omnipotent God? And in a God angered by human sin? Scientists have recently come up with a fascinating theory. To understand it, we have to backtrack to Chapter 3, where we learned that there is something unique about *Homo puppy*'s eyes. Thanks to the whites surrounding our irises, we can follow the direction of one another's gazes. The glimpse this gives us into other people's minds is vital to forging bonds of trust.

When we began living together in large groups alongside thousands of strangers, everything changed. Quite literally, we lost sight of each other. There's no way you can make eye contact with thousands or tens of thousands or a million people, so our mutual distrust began to grow. Increasingly, people started to suspect others of sponging off the community; that while they were breaking their backs all those others were putting their feet up.

And so rulers needed someone to keep tabs on the masses. Someone who heard everything and saw everything. An all-seeing Eye. God.

It's no accident that the new deities were vengeful types.[26] God became a super-Leviathan, spying on everyone twenty-four hours a day, seven days a week. Not even your thoughts were safe. 'Even the hairs of your head are all numbered,' the Bible tells us in Matthew 10:30. It was this omniscient being that from now on kept watch from the heavens, supervising, surveilling and – when necessary – striking.

Myths were key to helping the human race and our leaders do something no other species had done before. They enabled us to work together on a massive scale with millions of strangers. Furthermore, this theory goes on to say, it was from these great powers of fabrication that great civilisations arose. Judaism and Islam, nationalism and capitalism – all are products of our imagination. 'It all revolved around telling stories,' Israeli historian Yuval Noah Harari writes in his book *Sapiens* (2011), 'and convincing people to believe them.'[27]

This is a captivating theory, but it has one drawback.

It ignores 95 per cent of human history.

As it happens, our nomadic ancestors had already exceeded that magic threshold of one hundred and fifty friends.[28] Sure, we hunted and gathered in small groups, but groups also regularly swapped members, making us part of an immense network of cross-pollinating *Homo puppies*. We saw this in Chapter 3 with tribes like the Aché in Paraguay and the Hadza in Tanzania, whose members meet more than a thousand people over the course of their lifetimes.[29]

What's more, prehistoric people had rich imaginations, too. We've always spun ingenious myths that we passed down to each other and that greased the wheels of cooperation among multitudes. The world's earliest temple at Göbekli Tepe in modern-day Turkey (see Chapter 5) is a case in point, built through the concerted efforts of thousands.

The only difference is that in prehistory those myths were less stable. Chieftains could be summarily toppled and monuments speedily torn down. In the words of two anthropologists:

> Rather than idling in some primordial innocence, until the genie of inequality was somehow uncorked, our prehistoric ancestors seem to have successfully opened and shut the bottle on a regular basis, confining inequality to ritual costume dramas, constructing gods and kingdoms as they did their monuments, then cheerfully disassembling them once again.[30]

For millennia, we could afford to be sceptical about the stories we were told. If some loudmouth stood up announcing he'd been singled out by the hand of God, you could shrug it off. If that person became a nuisance, sooner or later he or she would get an arrow in the backside. *Homo puppy* was friendly, not naive.

It wasn't until the emergence of armies and their commanders that all this changed. Just try standing up to a strongman who has all opposition skinned, burned alive, or drawn and quartered. Your criticisms suddenly won't seem so urgent. 'This is the reason,' Machiavelli wrote, 'why all armed prophets have triumphed and all unarmed prophets have fallen.'

From this point on, gods and kings were no longer so easily ousted. Not backing a myth could now prove fatal. If you believed in the wrong god, you kept it to yourself. If you believed the nation state was a foolish illusion, it could cost you your head. 'It is useful,' advised Machiavelli, 'to arrange matters so that when they no longer believe, they can be made to believe by force.'[31]

You might think violence isn't a big part of the equation any more – at least not in tidy democracies with their boring bureaucracy. But make no mistake: the *threat* of violence is still very much present, and it's pervasive.[32] It's the reason families with children can be kicked out of their homes for defaulting on mortgage payments. It's the reason why immigrants can't simply stroll across the border in the fictions we call 'Europe' and 'the United States'. And it's also the reason we continue to believe in money.

Just consider: why would people hole up in cages we know as 'offices' for forty hours a week in exchange for some bits of metal and paper or a few digits added to their bank account? Is it because we've been won over by the propaganda of the powers that be? And, if so, why are there virtually no dissenters? Why does no one walk up to the tax authorities and say, 'Hey mister, I just read an interesting book about the power of myths and realised money is a fiction, so I'm skipping taxes this year.'

The reason is self-evident. If you ignore a bill or don't pay your taxes, you'll be fined or locked up. If you don't willingly comply, the authorities will come after you. Money may be a fiction, but it's enforced by the threat of very real violence.[33]

3

As I read Dacher Keltner's work and about the psychology of power, I began to see how the development of private property and farming could have led *Homo puppy* astray.

For millennia, we picked the nice guys to be in charge. We were well aware even in our prehistoric days that power corrupts, so we also leveraged a system of shaming and peer pressure to keep group members in check.

But 10,000 years ago it became substantially more difficult to unseat the powerful. As we settled down in cities and states and our rulers gained command over whole armies, a little gossip or a well-aimed spear were no longer enough. Kings simply didn't allow themselves to be dethroned. Presidents were not brought down by taunts and jeers.

Some historians suspect that we're now actually dependent on inequality. Yuval Noah Harari, for example, writes that 'complex human societies seem to require imagined hierarchies and unjust discrimination'.[34] (And you can be sure that such statements are met with grave approval at the top.)

But what fascinates me is that people around the world have continued to find ways to tame their leaders, even after the advent of chieftains and kings. One obvious method is revolution. Every revolution, whether the French (1789), the Russian (1917), or the Arab Spring (2011), is fuelled by the same dynamic. The masses try to overthrow a tyrant.

Most revolutions ultimately fail, though. No sooner is one despot brought down than a new leader stands up and develops an insatiable lust for power. After the French Revolution it was Napoleon. After the Russian Revolution it was Lenin and Stalin. Egypt, too, has reverted to yet another dictator. Sociologists call this the 'iron law of oligarchy': even socialists and communists, for all their vaunted ideals of liberty and equality, are far from immune to the corrupting influence of too much power.

Some societies have coped with this by engineering a system of distributed power – otherwise known as 'democracy'. Although the word suggests it is the people who govern (in ancient Greek, *demos* means 'people' and *kratos* means 'power'), it doesn't usually work out that way.

Rousseau already observed that this form of government is more accurately an 'elective aristocracy' because in practice the people are not in power at all. Instead we're allowed to decide who holds power over us. It's also important to realise this model was originally designed to exclude society's rank and file. Take the American Constitution: historians agree it 'was intrinsically an aristocratic document designed to check the democratic tendencies of the period'.[35] It was never the American Founding Fathers' intention for the general populace to play an active role in politics. Even now, though any citizen can run for public office, it's tough to win an election without access to an aristocratic network of donors and lobbyists. It's not surprising that American 'democracy' exhibits dynastic tendencies – think of the Kennedys, the Clintons, the Bushes.

Time and again we hope for better leaders, but all too often those hopes are dashed. The reason, says Professor Keltner, is that power causes people to lose the kindness and modesty that got them elected, or they never possessed those sterling qualities in the first place. In a hierarchically organised society, the Machiavellis are one step ahead. They have the ultimate secret weapon to defeat their competition.

They're shameless.

We saw earlier that *Homo puppy* evolved to experience shame. There's a reason that, of all the species in the animal kingdom, we're one of the few that blush. For millennia, shaming was the surest way to tame our leaders, and it can still work today. Shame is more effective than rules and regulations or censure and coercion, because people who feel shame regulate themselves. In the way their speech falters when they disappoint expectations or in a telltale flush when they realise they're the subject of gossip.[36]

Clearly, shame also has a dark side (shame induced by poverty, for example), but try to imagine what society would be like if shame didn't exist. That would be hell.

Unfortunately, there are always people who are unable to feel shame, whether because they are drugged on power or are among the small minority born with sociopathological traits. Such individuals wouldn't last long in nomadic tribes. They'd be cast out of the group and left to die alone. But in our modern sprawling organisations, sociopaths actually seem to be a few steps ahead on the career ladder. Studies show that between 4 and 8 per cent of CEOs have a diagnosable sociopathy, compared to 1 per cent among the general population.[37]

In our modern democracy, shamelessness can be positively advantageous. Politicians who aren't hindered by shame are free to do things others wouldn't dare. Would you call yourself your country's most brilliant thinker, or boast about your sexual prowess? Could you get caught in a lie and then tell another without missing a beat? Most people would be consumed by shame – just as most people leave that last cookie on the plate. But the shameless couldn't care less. And their audacious behaviour pays dividends in our modern mediacracies, because the news spotlights the abnormal and the absurd.

In this type of world, it's not the friendliest and most empathic leaders who rise to the top, but their opposites. In this world, it's survival of the shameless.

12

What the Enlightenment Got Wrong

I

After my foray into the psychology of power, my thoughts returned to the story in the Prologue of this book. It struck me that, in essence, the lessons of the previous chapters could all be found in that tale of the Blitz, of what transpired in London when the bombs fell.

The British authorities had predicted widespread panic. Looting. Riots. This kind of calamity would surely set off our inner brutes, hurling us into a war of all against all. But the opposite turned out to be true. Disasters bring out the best in us. It's as if they flip a collective reset switch and we revert to our better selves.

The second lesson of the Blitz is that we're groupish animals. Londoners supposed that their courage under fire was quintessentially British. They thought their resilience was akin to their stiff upper lip or dry sense of humour – just another element of a superior culture. In Chapter 10, we saw that this kind of group bias is typical of humans. We're all too inclined to think in terms of 'us' and 'them'. The tragedy of war is that it's the best facets of human nature – loyalty, camaraderie, solidarity – that inspire *Homo puppy* to take up arms.

Once we arrive at the front lines, however, we often lose our bluster. In Chapters 4 and 10 we saw that humans have a deep-rooted aversion to violence. For centuries, many soldiers couldn't even bring themselves to pull the trigger. Bayonets went unused. Most casualties were inflicted at a distance, by pilots or gunners who never needed to look their enemies in the eyes. This was also a lesson of the Blitz, where the worst assaults came from up high.

When the British then planned a bombing campaign of their own, the corrupting influence of power reared its ugly head. Frederick Lindemann, one of Churchill's inner circle, cast aside all evidence that bombs don't break morale. He'd already decided the Germans would cave in, and anyone who cared to contradict him was branded a traitor.

'The fact that the bombing policy was forced through with so little opposition,' remarked a later historian, 'is a typical example of the hypnosis of power.'[1]

And that finally brings us to an answer for that question posed by Hobbes and Rousseau. The question whether human nature is essentially good – or bad.

There are two sides to this answer, because *Homo puppy* is a thoroughly paradoxical creature. To begin with, we are one of the friendliest species in the animal kingdom. For most of our past, we inhabited an egalitarian world without kings or aristocrats, presidents or CEOs. Occasionally individuals did rise to power, but, as we saw in Chapter 11, they were brought down just as fast.

Our instinctive wariness of strangers posed no big problems for a long time. We knew our friends' names and faces, and if we crossed paths with a stranger we easily found common ground. There was no advertising or propaganda, no news or war that placed people in opposition. We were free to leave one group and join another, in the process building extended relational networks.

But then, 10,000 years ago, the trouble began.

From the moment we began settling in one place and amassing private property, our group instinct was no longer so innocuous. Combined with scarcity and hierarchies, it became downright toxic. And once leaders began raising armies to do their bidding, there was no stopping the corruptive effects of power.

In this new world of farmers and fighters, cities and states, we straddled an uncomfortably thin line between friendliness and xenophobia. Yearning for a sense of belonging, we were quickly inclined to repel outsiders. We found it difficult to say no to our own leaders – even if they marched us onto the wrong side of history.

With the dawn of civilisation, *Homo puppy*'s ugliest side came to the fore. History books chronicle countless massacres by Israelites and Romans, Huns and Vandals, Catholics and Protestants, and many more. The names change, but the mechanism stays the same: inspired by fellowship and incited by cynical strongmen, people will do the most horrific things to each other.

This has been our predicament for millennia. You could even see the history of civilisation as an epic struggle against the biggest mistake of all time. *Homo puppy* is an animal that has been wrenched from its natural habitat. An animal that has been turning itself inside out to bridge a cavernous 'mismatch' ever since. For thousands of years, we've been striving to exorcise the curse of disease, war and oppression I wrote about in Chapter 5 – to lift the curse of civilisation.

And then only recently, it looked as if we just might do it.

2

In the early seventeenth century, a movement began that we now call 'the Enlightenment'. It was a philosophical revolution.

The thinkers of the Enlightenment laid the foundations for the modern world, from the rule of law to democracy and from education to science.

At first glance, Enlightenment thinkers like Thomas Hobbes seemed not so different from earlier priests and ministers. All of them operated on the same assumption that human nature is corrupt. Scottish philosopher David Hume summarised the Enlightenment view as 'every man ought to be supposed a knave, and to have no other end, in all his actions, than private interest'.[2]

And yet, according to these thinkers, there was a way we could productively harness our self-interest. We humans have one phenomenal talent, they said, a saving grace that sets us apart from other living creatures. It's this gift that we could cling to. This was the miracle on which we might pin our hopes.

Reason.

Not empathy, or emotion, or faith. Reason. If Enlightenment philosophers put their faith in something, it was in the power of rational thought. They became convinced that humans could design intelligent institutions which factored in our innate selfishness. They believed we can paint a civilising layer over our darker instincts. Or, more precisely, that we could enlist our bad qualities to serve the common good.

If there was one sin that Enlightenment thinkers espoused, it was greed, which they trumpeted under the motto 'private vices, public benefits'.[3] This stood for the ingenious notion that behaviour which was antisocial at the individual level could have payoffs for wider society. Enlightenment economist Adam Smith set out this idea in his classic *The Wealth of Nations* (1776), which was the first book to defend the principles of the free market. In it, he famously wrote: 'It is not from the benevolence of the butcher, the brewer, or the baker that we expect our dinner, but from their regard to their own interest. We address ourselves, not to their humanity, but to their self-love, and never talk to them of our own necessities, but of their advantages.'

Selfishness should not be tamped down, modern economists argued, but turned loose. In this way, the desire for wealth would achieve what no army of preachers ever could: unite people the world over. Nowadays, when we pay for our groceries at the supermarket, we're working together with thousands of people who contribute to the production and distribution of the stuff in our trollies. Not out of the goodness of our hearts, but because we're looking out for ourselves.

Enlightenment thinkers used the same principle to underpin their model of modern democracy. Take the US Constitution, which is the world's oldest still in effect. Drawn up by the Founding Fathers, it is premised on their pessimistic view that our essentially selfish nature needs to be restrained. To this end, they set out a system of 'checks and balances', in which everybody kept an eye on everybody else.

The idea is that if those in power (from right to left, Republicans and Democrats), across top government institutions (the Senate and the House of Representatives, the White House and the Supreme Court), keep each other in check, then the American people will be able to live together in harmony despite their corrupt nature.[4] And the only way to rein in corruptible politicians, these rationalists believed, was by balancing them against other politicians. In the words of American statesman James Madison: 'Ambition must be made to counteract ambition.'

Meanwhile, this era also witnessed the birth of modern rule of law. Here was another antidote to our darker instincts, because Lady Justice is by definition blind. Unencumbered by empathy, love, or bias of any kind, justice is governed by reason alone. Likewise, it was reason that provided the underpinnings of our new bureaucratic systems, which subjected one and all to the same procedures, rules and laws.

From now on, you could do business with anyone you'd like, no matter their religion or creed. A side effect was that in those very countries with a strong rule of law, assuring regulations and contracts would be honoured, belief in a vengeful God diminished. The role of God the father was to be supplanted by faith in the state. In the wake of the Enlightenment, religion consequently adopted a much friendlier demeanour. These days, few states still defer to the judging eye of God, and instead of calling for bloody crusades, popes give heartwarming speeches about 'a revolution of tenderness'.[5]

Can it be a coincidence that the largest concentrations of atheists are to be found in countries like Denmark or Sweden? These nations also have the most robust rule of law and most trustworthy bureaucracies.[6] In countries like these, religion has been displaced. Much as mass production once sidelined traditional craftspeople, God lost his job to bureaucrats.

So here we are, a few centuries into the Age of Reason. All things considered, we have to conclude that the Enlightenment has been a triumph for humankind, bringing us capitalism, democracy and the rule of law. The statistics are clear. Our lives are exponentially better and the world is richer, safer and healthier than ever before.[7]

Only two hundred years ago, any kind of settled life still meant extreme poverty, no matter where in the world you lived. These days, that's true for less than 10 per cent of the global population. We have as good as conquered the biggest infectious diseases, and even if the news might lead you to think otherwise, the last few decades have seen rates for everything from child mortality and starvation to murders and war casualties plummet spectacularly.[8]

So how can we live harmoniously if we distrust strangers? How can we exorcise the curse of civilisation, disease, slavery and the oppression that plagued us for 10,000 years? The cold, hard reason of the Enlightenment provided an answer to this old dilemma.

And it was the best answer – until now.

Because, let's be honest, the Enlightenment also had a dark side. Over the past few centuries we've learned that capitalism can run amok, sociopaths can seize power and a society dominated by rules and protocols has little regard for the individual.

Historians point out that if the Enlightenment gave us equality, it also invented racism. Eighteenth-century philosophers were the first to classify humans into disparate 'races'. David Hume, for instance, wrote that he was 'apt to suspect the negroes [...] to be naturally inferior to the whites'. In France, Voltaire agreed: 'If their understanding is not of a different nature from ours, it is at least greatly inferior.' Such racist ideas became encoded in legislation and norms of conduct. Thomas Jefferson, who penned the immortal words 'all men are created equal' in the American Declaration of Independence, was a slave owner. He also said: 'Never yet could I find that a black had uttered a thought above the level of plain narration.'

And then came the bloodiest conflict in history. The Holocaust unfolded in what had been a cradle of the Enlightenment. It was effectuated by an ultra-modern bureaucracy, in which management of the concentration camps was tasked to the SS Main 'Economic and Administrative' Department. Many scholars have thus come to regard the extermination of six million Jews as not only the height of brutality, but of modernity.[9]

The Enlightenment's contradictions stand out when we examine its portrayal of human nature. On the face of it, philosophers like David Hume and Adam Smith took a cynical view. Modern capitalism, democracy and the rule of law are all founded on the principle that people are selfish. But if you actually read their books you come to realise that Enlightenment authors were not diehard cynics at all. Seventeen years before publishing *The Wealth of Nations* (destined to become the capitalist bible), Adam Smith wrote a volume titled *The Theory of Moral Sentiments*. In it, we find passages like this one:

> How selfish soever man may be supposed, there are evidently some principles in his nature, which interest him in the fortune of others, and render their happiness necessary to him, though he derives nothing from it except the pleasure of seeing it.

Influential rationalists like Smith and Hume made a point of emphasising the vast capacity humans show for empathy and altruism. Why then, if all these philosophers were so attuned to our admirable qualities, were their institutions (democracy, trade and industry) so often premised on pessimism? Why did they continue to cultivate a negative view of human nature?

We can trace the answer in one of David Hume's books, in which the Scottish philosopher articulates precisely this contradiction in Enlightenment thought:

'It is, therefore, a just political maxim, that every man must be supposed a knave: though at the same time, it appears somewhat strange, that a maxim should be true in politics, which is false in fact.'

In other words, Hume believed that we should act *as though* people have a selfish nature. Even though we know they don't. When I realised this, a single word flashed through my mind: *nocebo*. Could this be the thing that the Enlightenment – and, by extension, our modern society – gets wrong? That we continually operate on a mistaken model of human nature?

In Chapter 1 we saw that some things can become true merely because we believe in them – that pessimism becomes a self-fulfilling prophecy. When modern economists assumed that people are innately selfish, they advocated policies that fostered self-serving behaviour. When politicians convinced themselves that politics is a cynical game, that's exactly what it became.

So now we have to ask: could things be different?

Can we use our heads and harness rationality to design new institutions? Institutions that operate on a wholly different view of human nature? What if schools and businesses, cities and nations expect the best of people instead of presuming the worst?

These questions are the focus of the rest of this book.

Part Four

A NEW REALISM

'So we have to be idealists in a way, because then we wind up
as the true, the real realists.'

Viktor Frankl (1905–97)

I

I was nineteen when I attended my first lecture in philosophy. That morning, sitting under the bright fluorescent lights of an auditorium at Utrecht University, I made the acquaintance of British mathematician and philosopher Bertrand Russell (1872–1970). There and then, he became my new hero.

Besides being a brilliant logician and the founder of a revolutionary school, Russell was an early advocate for homosexuals, a freethinker who foresaw the Russian Revolution ending in misery, an anti-war activist who would be thrown behind bars for civil disobedience at the age of eighty-nine, author of more than sixty books and two thousand articles, and the survivor an aircraft crash. He also won the Nobel Prize for Literature.

What I personally admired most about Russell was his intellectual integrity, his fidelity to truth. Russell understood the all-too-human proclivity to believe what suits us, and he resisted it his whole life. Time and again, he swam against the tide, knowing it would cost him dearly. One statement of his particularly stands out for me. In 1959, the BBC asked Russell what advice he would give future generations. He answered:

> When you are studying any matter or considering any philosophy, ask yourself only what are the facts and what is the

truth that the facts bear out. Never let yourself be diverted either by what you wish to believe or by what you think would have beneficent social effects if it were believed, but look only and solely at what are the facts.

These words had a big impact on me. They came at a time when I'd begun to question my own faith in God. As a preacher's son and member of a Christian student society, my instinct was to cast my doubts to the wind. I knew what I wanted: I wanted there to be life after death, for all the world's wrongs to be made right in the hereafter and for us not to be on our own on this rock in the universe.

But from then on I would be haunted by Russell's warning: 'Never let yourself be diverted by what you wish to believe.'

I've done my best, while writing this book.

Did I succeed in following Russell's advice? I hope so. At the same time, I have my doubts. I know I needed a lot of help from critical readers to keep me on track. But then, to quote Russell himself, 'None of our beliefs are quite true; all have at least a penumbra of vagueness and error.' So if we aim to get as close as possible to the truth, we have to eschew certainty and question ourselves every step of the way. 'The Will to Doubt', Russell called this approach.

It wasn't until years after learning about this British thinker that I discovered his maxim contained a reference. Russell coined the phrase 'The Will to Doubt' to place himself in opposition to another philosopher, an American named William James (1842–1910).

And this is who I want to tell you about now. William James was the mentor of Theodore Roosevelt, Gertrude Stein, W. E. B. Du Bois, and many other leading lights of American history.

He was a beloved figure. According to Russell, who'd met him, James was 'full of the warmth of human kindness'.

Yet Russell was less enamoured of James's ideas. In 1896, he had delivered a talk not about the will to doubt, but 'The Will to Believe'. James professed that some things just have to be taken on faith, even if we can't prove they're true.

Take friendship. If you go around for ever doubting other people, you'll behave in ways guaranteed to make you disliked. Things like friendship, love, trust and loyalty become true precisely *because* we believe in them. While James allowed that one's belief could be proved wrong, he argued that 'dupery through hope' was preferable to 'dupery through fear'.

Bertrand Russell didn't go in for this kind of mental gymnastics. Much as he liked the man himself, he disliked James's philosophy. The truth, he said, doesn't deal in wishful thinking. For many years that was my motto, too — until I began to doubt doubt itself.

2

The year is 1963, four years after Russell's interview with the BBC.

In Cambridge, Massachusetts, the young psychologist Bob Rosenthal decides to try a little experiment at his Harvard University lab. Beside two rat cages, he posts different signs identifying the rodents in one as specially trained intelligent specimens and in the other as dull and dim-witted.

Later that day, Rosenthal instructs his students to put the rats in a maze and record how long it takes each one to find its way out. What he doesn't tell the students is that in fact none of the animals is special in any way — they're all just ordinary lab rats.

But then something peculiar happens. The rats that the students *believe* to be brighter and faster really do perform better.

It's like magic. The 'bright' rats, though no different from their 'dull' counterparts, perform twice as well.

At first, no one believed Rosenthal. 'I was having trouble publishing any of this,' he recalled decades later.[1] Even he had trouble initially accepting that there were no mysterious forces at play and that there's a perfectly rational explanation. What Rosenthal came to realise is that his students handled the 'bright' rats – the ones of which they had higher expectations – more warmly and gently. This treatment changed the rats' behaviour and enhanced their performance.

In the wake of his experiment, a radical idea took root in Rosenthal's mind; a conviction that he had discovered an invisible yet fundamental force. 'If rats became brighter when expected to,' Rosenthal speculated in the magazine *American Scientist*, 'then it should not be farfetched to think that children could become brighter when expected to by their teachers.'

A few weeks later, a letter arrived for the psychologist. It was from the principal of Spruce Elementary School in San Francisco, who had read Rosenthal's article and extended an irresistible offer. 'Please let me know whether I can be of assistance,' she wrote.[2] Rosenthal didn't need to think twice. He immediately set to work designing a new experiment. This time his subjects weren't lab rats, but children.

When the new school year started, teachers at Spruce Elementary learned that an acclaimed scientist by the name of Dr Rosenthal would be administering a test to their pupils. This 'Test of Inflected Acquisition' indicated who would make the greatest strides at school that year.

In truth it was a common or garden IQ test, and, once the scores had been tallied, Rosenthal and his team cast them all aside. They tossed a coin to decide which kids they would tell teachers were 'high-potentials'. The kids, meanwhile, were told nothing at all.

Sure enough, the power of expectation swiftly began to work its magic. Teachers gave the group of 'smart' pupils more attention, more encouragement and more praise, thus changing how the children saw themselves, too. The effect was clearest among the youngest kids, whose IQ scores increased by an average of twenty-seven points in a single year. The largest gains were among boys who looked Latino, a group typically subject to the lowest expectations in California.[3]

Rosenthal dubbed his discovery the Pygmalion Effect, after the mythological sculptor who fell so hard for one of his own creations that the gods decided to bring his statue to life. Beliefs we're devoted to – whether they're true or imagined – can likewise come to life, effecting very real change in the world. The Pygmalion Effect resembles the placebo effect (which I discussed in Chapter 1), except, instead of benefiting oneself, these are expectations that benefit others.

At first I thought a study this old would surely have been debunked by now, like all those other mediagenic experiments from the 1960s.

Not at all. Fifty years on, the Pygmalion Effect remains an important finding in psychological research. It's been tested by hundreds of studies in the army, at universities, in courtrooms, in families, in nursing homes and within organisations.[4] True, the effect isn't always as strong as Rosenthal initially thought, especially when it comes to how children perform on IQ tests. Nonetheless, a critical review study in 2005 concluded that 'the abundant naturalistic and experimental evidence shows that teacher expectations clearly do influence students – at least sometimes'.[5] High expectations can be a powerful tool. When wielded by managers, employees perform better. When wielded by officers, soldiers fight harder. When wielded by nurses, patients recover faster.

Despite this, Rosenthal's discovery didn't spark the revolution he and his team had hoped for. 'The Pygmalion Effect is great science that is underapplied,' an Israeli psychologist has lamented. 'It hasn't made the difference it should have in the world, and that's very disappointing.'[6]

I've got more bad news: just as positive expectations have very real effects, nightmares can come true, too. The flip side of the Pygmalion Effect is what's known as the Golem Effect, named after the Jewish legend in which a creature meant to protect the citizens of Prague instead turns into a monster. Like the Pygmalion Effect, the Golem Effect is ubiquitous. When we have negative expectations about someone, we don't look at them as often. We distance ourselves from them. We don't smile at them as much. Basically, we do exactly what Rosenthal's students did when they released the 'stupid' rats into the maze.

Research on the Golem Effect is scant, which is not surprising, given the ethical objections to subjecting people to negative expectations. But what we do know is shocking. Take the study done by psychologist Wendell Johnson in Davenport, Iowa, in 1939. He split twenty orphans up into two groups, telling one that they were good, articulate speakers and the other that they were destined to become stutterers. Now infamously known as 'The Monster Study', this experiment left multiple individuals with lifelong speech impediments.[7]

The Golem Effect is a kind of nocebo: a nocebo that causes poor pupils to fall further behind, the homeless to lose hope and isolated teenagers to radicalise. It's also one of the insidious mechanisms behind racism, because when you're subjected to low expectations, you won't perform at your best, which further diminishes others' expectations and thus further undermines your performance. There's also evidence to suggest that the Golem Effect and its vicious cycle of mounting negative expectations can run entire organisations into the ground.[8]

3

The Pygmalion and Golem Effects are woven into the fabric of our world. Every day, we make each other smarter or stupider, stronger or weaker, faster or slower. We can't help leaking expectations, through our gazes, our body language and our voices. My expectations about you define my attitude towards you, and the way I behave towards you in turn influences your expectations and therefore your behaviour towards me.

If you think about it, this gets to the very crux of the human condition. *Homo puppy* is like an antenna, constantly attuned to other people. Somebody else's finger gets trapped in the door and you flinch. A tightrope walker balances on a thin cord and you feel your own stomach lurch. Someone yawns and it's almost impossible for you not to yawn as well. We're hardwired to mirror one another.

Most of the time, this mirroring works well. It fosters connections and good vibes, as when everybody's grooving together on the dance floor. Our natural instinct to mirror others tends to be seen in a positive light for precisely this reason, but the instinct works two ways. We also mirror negative emotions such as hatred, envy and greed.[9] And when we adopt one another's bad ideas — thinking them to be ideas everybody around us holds — the results can be downright disastrous.

Take economic bubbles. Back in 1936, British economist John Maynard Keynes concluded that there was a striking parallel between financial markets and beauty pageants. Imagine you're presented with a hundred contestants, but, rather than picking your own favourite, you have to indicate which one *others* will prefer.[10] In this kind of situation, our inclination is to guess what other people will think. Likewise, if everybody thinks everybody else thinks that the value of a share will go up, then the share value goes up. This can go on for a long time, but eventually the bubble bursts. That happened, for example, when tulip mania hit Holland in January 1637, and a single tulip bulb briefly sold for more than ten times the annual wage of a skilled craftsman, only to become all but worthless days later.

Bubbles of this kind are not isolated to the financial world. They're everywhere. Dan Ariely, a psychologist at Duke University, once gave a brilliant demonstration during a college lecture. To explain his field of behavioural economics, he provided the class with what sounded like an extremely technical definition. Unbeknown to the students, however, all the terms he used had been generated by a computer, cobbled together in a series of random words and sentences to produce gibberish about 'dialectic enigmatic theory' and 'neodeconstructive rationalism'.

Ariely's students – at one of the world's top universities – listened with rapt attention to this linguistic mash-up. Minutes ticked by. No one laughed. No one raised their hand. No one gave any sign they didn't understand.

'And this brings us to the big question ...' Ariely finally concluded. 'Why has no one asked me what the #$?@! I'm talking about?'[11]

In psychology circles, what happened in that classroom is known as *pluralistic ignorance* – and, no, this isn't a term generated by a machine. Individually, Ariely's students found his narrative impossible to follow, but because they saw their classmates listening attentively, they assumed the problem was them. (This phenomenon is no doubt familiar to readers who have attended conferences on topics like 'disruptive co-creation in the network society'.)

Though harmless in this instance, research shows that the effects of pluralistic ignorance can be disastrous – even fatal. Consider binge drinking. Survey college students on their own, and most will say drinking themselves into oblivion isn't their favourite pastime. But because they assume *other* students are big fans of drinking, they try to keep up and everyone winds up puking in the gutter.

Researchers have compiled reams of data demonstrating that this kind of negative spiral can also factor into deeper societal evils like racism, gang rape, honour killings, support for terrorists and dictatorial regimes, even genocide.[12] While condemning these acts in their own minds, perpetrators fear they're alone and therefore decide to go with the flow. After all, if there's one thing *Homo puppy* struggles with, it's standing up to the group. We prefer a pound of the worst kind of misery over a few ounces of shame or social discomfort.

This led me to wonder: what if our negative ideas about human nature are actually a form of pluralistic ignorance? Could our fear that most people are out to maximise their own gain be born of the assumption that that's what others think? And then we adopt a cynical view when, deep down, most of us are yearning for a life of more kindness and solidarity?

I'm reminded sometimes of how ants can get trapped crawling in circles. Ants are programmed to follow each other's pheromone trails. This usually results in neat trails of ants, but occasionally a group will get sidetracked and wind up 'travelling' in a continuous circle. Tens of thousands of ants can get trapped rotating in circles hundreds of feet wide. Blindly they carry on, until they succumb to exhaustion and lack of food and die.

Every now and then families, organisations, even entire countries seem to get caught in these kinds of spirals. We keep going around in circles, assuming the worst about each other. Few of us move to resist and so we march to our own downfall.

It's been fifty years since Bob Rosenthal's career began, and to this day he still wants to figure out how we can use the power of expectation to our advantage. Because he knows that, like hatred, trust can also be contagious.

Trust often begins when someone dares to go against the flow – someone who's initially seen as unrealistic, even naive. In the next part of this book, I want to introduce you to several such individuals. Managers who have total faith in their staff. Teachers who give kids free rein to play. And elected officials who treat their constituents as creative, engaged citizens.

These are people fuelled by what William James called 'The Will to Believe'. People who recreate the world in their own image.

13

The Power of Intrinsic Motivation

I

I'd been eager to meet Jos de Blok for some time. Having read about the success of his home healthcare organisation *Buurtzorg*, I had a hunch he was one of those exponents of a new realism. Of a new view of human nature.

But to be honest, the first time I talked to him he didn't strike me as a great thinker. In one sweeping statement he dismissed the whole management profession: 'Managing is bullshit. Just let people do their job.'

Sure, Jos, you think. Have another. But then you realise: this isn't some crackpot talking. This is a guy who's built a hugely successful organisation employing over fourteen thousand people. Who's been voted Employer of the Year in the Netherlands five times. Professors from New York to Tokyo travel all the way to the town of Almelo to witness his wisdom first-hand.

I went back to have a look at the interviews Jos de Blok has given. They soon had me grinning:

> Interviewer: Is there anything you do to motivate yourself? Steve Jobs reportedly asked himself in the mirror each morning: What would I do if this were my last day?
> Jos: I read his book too, and I don't believe a word of it.[1]
> Interviewer: Do you ever attend networking sessions?

Jos: At most of those things, nothing happens aside from everybody reaffirming everybody else's opinions. That's not for me.[2]

Interviewer: How do you motivate your employees?

Jos: I don't. Seems patronizing.[3]

Interviewer: What's your speck on the horizon, Jos – that distant goal that inspires you and your team?

Jos: I don't have distant goals. Not all that inspired by specks.[4]

Unlikely as it may sound, this is also a man who's received the prestigious Albert Medal from the Royal Society of Arts in London, ranking him with the likes of Tim Berners-Lee, brain behind the World Wide Web; Francis Crick, who unravelled the structure of DNA; and the brilliant physicist Stephen Hawking. In November 2014, it was Jos de Blok from small-town Holland who received the honour and the cream of British academia turned out to attend his keynote speech. In broken English, De Blok confessed that at first he thought it was a joke.

But it was no joke.

It was high time.

2

To understand what makes de Blok's ideas so revolutionary – and on par with cracking DNA – we have to go back to the beginning of the twentieth century. That's when business administration made its debut. This new field of science had its roots planted firmly in the Hobbesian view that human beings are greedy by nature. We needed managers to keep us on the straight and narrow. Managers – the thinking went – had to provide us with the right 'incentives'. Bankers get bonuses because it makes them work harder. Unemployment benefits are conditional to force people off the couch. Kids get F grades to make them put in a better effort next term.

What's fascinating is that the two major ideologies of the twentieth century – capitalism and communism – both shared this view of humanity. Both the capitalist and the communist would tell you that there are only two ways to propel people into action: carrots and sticks. The capitalists relied on carrots (read: money), whereas the communists were mainly about sticks (read: punishment). For all their differences, there was one basic premise on which both sides could agree: People don't motivate themselves.

Now you may be thinking: *Oh, it's not as bad as that. I, for one, am plenty motivated.*

I'm not going to argue. In fact, I'm sure you're right. My point is that we tend to think *those other people* lack motivation. Professor Chip Heath of Stanford University refers to this as our *extrinsic incentives bias*. That is, we go around assuming other people can be motivated solely by money. In a survey among law students, for instance, Heath discovered that 64 per cent said they were studying law because it was a long-time dream or because it interested them. Only 12 per cent believed the same held true for their peers. All those other students? They were in it for the money.[5]

It's this cynical view of humankind that laid the foundations for capitalism. 'What workers want most from their employers, beyond anything else, is high wages,' asserted one of the world's first business consultants, Frederick Taylor, some hundred years ago.[6] Taylor made his name as the inventor of *scientific management*, a method premised on the notion that performance must be measured with the greatest possible precision in order to make factories as efficient as possible. Managers had to be stationed at every production line, stopwatch at the ready, to record how long it took to tighten a screw or pack a box. Taylor himself likened the ideal employee to a brainless robot: 'so stupid and so phlegmatic that he more nearly resembles an ox.'[7]

With this cheery message, Frederick Taylor grew to become one of the most renowned management scientists ever. In the early twentieth century, the whole world was giddy with his ideas – communists, fascists and capitalists alike. From Lenin to Mussolini, from Renault to Siemens – Taylor's management philosophy continued to spread. In the words of his biographer, Taylorism 'adapts the way a virus does, fitting in almost everywhere'.[8]

Of course, a lot has changed since Taylor's day. You'll now find plenty of start-ups where you can show up at the office wearing flip flops. And many workers these days have the flexibility to set their own hours. But Taylor's view of humanity, and the conviction that only carrots and sticks get people moving, is as pervasive as ever. Taylorism lives on in timesheets, billable hours and KPIs, in doctor pay-for-performance programmes and in warehouse staff whose every move is monitored by CCTV.

3

The first murmurs of dissent came in the summer of 1969.

Edward Deci was a young psychologist working on his Ph.D. at a time when the field was in a thrall to *behaviourism*. This theory held – like Frederick Taylor's – that people are shiftless creatures. The only thing powerful enough to spur us to action is the promise of a reward or the fear of punishment.

Yet Deci had a nagging sense that this theory didn't stack up. After all, people go around doing all kinds of nutty things that don't fit the behaviourist view. Like climbing mountains (hard!), volunteering (free!) and having babies (intense!). In fact, we're continually engaging in activities – of our own free will – that don't earn us a penny and are downright exhausting. Why?

That summer Deci stumbled across a strange anomaly: in some cases, carrots and sticks can cause performance to *slack off*. When he paid student volunteers a dollar to solve a puzzle, they lost interest in the task. 'Money may work', Deci later explained, 'to buy off one's intrinsic motivation for an activity.'[9]

This hypothesis was so revolutionary that economists rejected it out of hand. Financial incentives only served to increase motivation, they steadfastly maintained, so if a student enjoyed doing a puzzle, a reward would make her even more enthusiastic. Fellow psychologists were equally contemptuous of Edward Deci and his ideas. 'We were out of the mainstream,' recalled his co-researcher and closest friend Richard Ryan. 'The idea that rewards would sometimes undermine motivation was anathema to behaviorists.'[10]

But then a steady stream of studies began corroborating Deci's suspicions. Take the one done in Haifa, Israel, in the late 1990s, where a chain of day care centres found itself in a predicament. A quarter of the parents picked their kids up late, arriving past closing time. The result was fussy children and staff forced to work overtime. And so the organisation decided to impose a late fee: three dollars every time a parent showed up late.

Sounds like a good plan, right? Now parents had not one but two incentives – both a moral and a financial one – to arrive on time.

The new policy was announced, and the number of parents who arrived late ... went up. Before long, one-third was arriving after closing time, and in a matter of weeks, 40 per cent. The reason was straightforward: parents interpreted the late fee not as a fine but as a surcharge, which now released them from their obligation to pick up their kids on time.[11]

Many other studies have since validated this finding. It turns out that, under some conditions, the reasons people do things can't all be added up together. Sometimes, they'll cancel each other out.

A few years ago, researchers at the University of Massachusetts analysed fifty-one studies on the effects of economic incentives in the workplace. They found 'overwhelming evidence' that bonuses can blunt the intrinsic motivation and moral compass of employees.[12] And as if that wasn't bad enough, they also discovered that bonuses and targets can erode creativity. Extrinsic incentives will generally pay out in kind. Pay by the hour and you get more hours. Pay by the publication and you get more publications. Pay by the surgical procedure and you get more surgical procedures.

Here again, the parallels between the western capitalist economy and that of the former Soviet Union are striking. Soviet-era managers worked with targets. When targets went up – say, at a furniture factory – the quality of the furniture plummeted. Next when it was decided tables and chairs would be priced by weight, suddenly the factory would produce pieces too heavy to move.

This may sound amusing, but the sad truth is that it's still happening in many organisations today. Surgeons paid on a per-treatment basis are more inclined to sharpen their scalpels than to deliver better care. A big law firm that requires its staff to bill a minimum number of hours (say 1,500 a year) isn't stimulating its lawyers to work better, only longer. Communist or capitalist, in both systems the tyranny of numbers drowns out our intrinsic motivation.

So are bonuses a complete waste of money? Not entirely. Research by behavioural economist Dan Ariely has shown they can be effective when the tasks are simple and routine, like those Frederick Taylor timed on his stopwatch on production lines.[13] Precisely the kinds of tasks, in other words, that modern economies increasingly get robots to do, and robots have no need for intrinsic motivation.

But we humans can't do without.

Sadly, the lessons of Edward Deci haven't made it into daily practice nearly enough. Too often people are still treated like robots. At the office. At school. In hospital. At social services.

Time and again, we assume that others care only about themselves. That, unless there's a reward in the offing, people much prefer to lounge around. A British study recently found that a vast majority of the population (74 per cent) identify more closely with values such as helpfulness, honesty and justice than with wealth, status and power. But just about as large a share (78 per cent) think *others* are more self-interested than they really are.[14]

Some economists think this skewed take on human nature isn't a problem. Nobel Prize-winning economist Milton Friedman, for instance, argued that incorrect assumptions about people don't matter so long as your predictions prove right.[15] But Friedman forgot to factor in the nocebo effect: simply believing in something can make it come true.

How you get paid for what you do can turn you into an entirely different person. Two American psychologists demonstrated a few years ago that lawyers and consultants who are paid by the hour put a price on all their time, even outside the office. The upshot? Lawyers who meticulously log their hours are also less inclined to do pro bono work.[16]

It's mind-boggling to see how we get tripped up by targets, bonuses and the prospect of penalties:

- Think about CEOs who focus solely on quarterly results, and wind up driving their companies into the ground.

- Academics who are evaluated on their published output, and then tempted to put forward bogus research.
- Schools that are assessed on their standardised test results, and so skip teaching those skills that can't be quantified.
- Psychologists who are paid to continue to treat patients, and thus keep patients in treatment longer than necessary.
- Bankers who earn bonuses by selling subprime mortgages, and end up bringing the global economy to the brink of ruin.

The list continues. A hundred years after Frederick Taylor, we're still busy undermining one another's intrinsic motivation on a massive scale. A major study among 230,000 people across 142 countries revealed that a mere 13 per cent actually feel 'engaged' at work.[17] Thirteen per cent. When you wrap your brain around these kinds of figures, you realise how much ambition and energy are going to waste.

And how much room there is to do things differently.

4

Which brings us back to Jos de Blok. Up until early 2006, he sat on the board of directors at a large Dutch healthcare organisation. He floated one idea after another for 'self-directed teams' and 'hands-off management' until he had his fellow managers seeing red. De Blok himself had no business training or degree. He'd started studying economics years before only to drop out and become a nurse.

'The gap between the people at the top and the folks doing the actual work – in healthcare, in education, you name it – is enormous,' De Blok tells me when I visit his office in Almelo. 'Managers tend to band together. They set up all kinds of courses and conferences where they tell each other they're doing things right.'

That cuts them off from the real world. 'There's this notion that doers can't think strategically,' De Blok continues. 'That they lack vision. But the people out doing the work are brimming with ideas. They come up with a thousand things, but don't get heard, because managers think they have to go on some corporate retreat to dream up plans to present to the worker bees.'

De Blok has a very different take on things. He sees his employees as intrinsically motivated professionals and experts on how their jobs ought to be done. 'In my experience, managers tend to have very few ideas. They get their jobs because they fit into a system, because they follow orders. Not because they're big visionaries. They take some "high-performance leadership" courses and suddenly think they're a game changer, an innovator.'

When I point out to de Blok that 'healthcare manager' was the fastest-growing occupational group in the Netherlands in the years 1996–2014, he heaves a sigh.[18]

'What you get with all these MBA programmes is people convinced they've learned a convenient way to order the world. You have HR, finance, IT. Eventually, you start believing that a lot of what your organisation is accomplishing is down to you. You see it with loads of managers. But subtract management and the work continues as before – or even better.'

As statements like this testify, de Blok tends to swim against the tide.

He's a manager who prefers not to manage. A CEO with hands-on experience. An anarchist at the top of the ladder. So as care became a product and patients became customers, De Blok decided to give up his management job and start something new. He dreamed of an oasis in this vast bureaucratic wasteland, a place fuelled not by market forces and growth, but by small teams and trust.

Buurtzorg started out with one team of four nurses in Enschede, a Dutch city of 150,000 on the country's eastern fringes. Today, it numbers more than eight hundred teams active nationwide. However, it's not what the organisation is, but what it is *not*, that sets Buurtzorg apart. It has no managers, no call centre and no planners. There are no targets or bonuses. Overheads are negligible and so is time spent in meetings. Buurtzorg doesn't have a flashy HQ in the capital, but occupies an uninspiring block in an ugly business park in outlying Almelo.

Each team of twelve has maximum autonomy. Teams plan their own schedules and employ their own co-workers. And, unlike the rest of the country's infinitely scripted care industry, the teams don't supply code H126 ('Personal Care'), code H127 ('Additional Personal Care'), code H120 ('Special Personal Care'), or code H136 ('Supplemental Remote Personal Care'). No, Buurtzorg supplies just one thing: care. In the exhaustive 'Product Book' of 'Care Products' defined by insurers, Buurtzorg now has its very own code: R002 – 'Buurtzorg'.

For the rest, the organisation has an intranet site where colleagues can pool their knowledge and experience. Each team has its own training budget, and each group of fifty teams has a coach they can call in if they get stuck. Finally, there's the main office which takes care of the financial side of things.

And that's it. With this simple formula, Buurtzorg has been proclaimed the country's 'Best Employer' five times, despite having no HR team, and won an award for 'Best Marketing in the Care Sector', despite having no marketing department. 'Employee and client satisfaction is phenomenally high,' concluded one consultant at KPMG. 'Though costing only slightly less than the average, their quality of care clearly exceeds the average.'[19]

That's right, Buurtzorg is better for patients, nicer for employees and cheaper for taxpayers. A win-win-win situation. Meanwhile, the organisation continues to grow. Every month, dozens of nurses leave other jobs to sign on with Buurtzorg. And no wonder: it gives them more freedom and more pay. When Buurtzorg recently acquired part of a bankrupt counterpart, de Blok announced: 'The first thing we're going to do is raise staff salaries.'[20]

Don't get me wrong: Buurtzorg isn't perfect. There are disagreements, things go awry – really, they're almost human. And the organisational structure is, if anything, old-fashioned, with de Blok's aim always having been a return to Holland's uncomplicated domestic healthcare services of the 1980s.

But the bottom line is that what Jos de Blok started back in 2006 is nothing short of extraordinary. You might say his organisation combines the best of left and right, spending taxpayer money on the delivery of small-scale care by independent practitioners.

De Blok sums up his philosophy like this: 'It's easy to make things hard, but hard to make them easy.' The record clearly shows that managers prefer the complicated. 'Because that makes your job more interesting,' de Blok explains. 'That lets you say: See, you need me to master that complexity.'

Could it be that's also driving a big part of our so-called 'knowledge economy'? That pedigree managers and consultants make simple things as complicated as possible so we will need them to steer us through all the complexity? Sometimes I secretly think this is the revenue model of not only Wall Street bankers but also postmodern philosophers peddling incomprehensible jargon. Both make simple things impossibly complex.

Jos de Blok does the opposite: he opts for simplicity. While healthcare conferences feature highly paid trend-watchers auguring disruption and innovation, he believes it's more important to preserve what works. 'The world benefits more from continuity than from continual change,' he asserts. 'Now they've got change managers, change agents, and so forth, but when I look at actual care in the community, the job has scarcely changed in thirty years. You need to build a relationship with someone in a tough situation; that's a constant. Sure, you may add some new insights and techniques, but the basics haven't changed.'

What does need to change, De Blok will tell you, is the care *system*. In recent decades, healthcare has been colonised by lawyers. 'Now you're in opposing camps. One side sells, the other buys. Just last week I was at a hospital where they told me: *We have our own sales team now.* It's crazy! We have hospitals with commercial departments and procurement teams, all staffing people with no background in healthcare at all. One buys, the other sells, and neither have a clue what it's all about.'

All the while, the bureaucracy keeps proliferating, because when you turn healthcare into a market, you end up with piles of paperwork. 'Nobody trusts anybody else, so they start building in all these safeguards; all kinds of checks that result in a ton of red tape. It's downright absurd,' says De Blok. 'The number of consultants and administrators at insurance companies is growing, while the number of actual caregivers continues to shrink.'

De Blok advocates a radically different approach to healthcare funding. Scrap the product mentality, he says. Make care central again. Drastically simplify the costs. 'The simpler the billing, the greater the emphasis on actual care,' he explains. 'The more complicated the billing, the more players will search for loopholes in the system, increasingly tipping the balance towards accounting departments until they're the ones defining care.'

Talking to Jos de Blok, it soon becomes clear that his lessons go beyond the care sector. They apply to other areas, too: to education and law enforcement, to government and industry.

A great example is FAVI, a French firm that supplies car parts. When Jean-François Zobrist was appointed its new CEO in 1983, FAVI had a rigid hierarchy structure and still did things the old-fashioned way. Work hard, you'll get a bonus. Clock in late, your wages will be docked.

From day one, Zobrist imagined an organisation in which not he, but his staff made the decisions. Where employees felt it their duty to arrive on time (and where you could be certain that if they didn't, they had a good reason). 'I dreamed of a place,' Zobrist recounted later, 'that everyone would treat like home. Nothing more, nothing less.'[21]

His first act as CEO was to brick up the huge window that let management keep an eye on the whole shop floor. Next he binned the time clock, had the locks taken off the storage rooms and axed the bonus system. Zobrist split the company into 'mini-factories' of twenty-five to thirty employees and had them each choose their own team leader. To these he gave free rein to make all their own decisions: on wages, working hours, who to hire, and all the rest. Each team answered directly to their customers.

Zobrist also decided not to replace the firm's old managers when they retired, and cut out the HR, planning and marketing departments. FAVI switched to a 'reverse delegation' method of working, in which teams did everything on their own, unless they themselves wanted to call in management.

This may sound like the recipe for a money-guzzling hippy commune, but in fact productivity at FAVI went *up*. The company workforce expanded from one hundred to five hundred and it went on to conquer 50 per cent of the market for transmission forks. Average production times for key parts dropped from eleven days to just one. And while competitors were forced to relocate operations to low-wage countries, the FAVI plant stayed put in Europe.[22]

All that time, Zobrist's philosophy was dead simple. If you treat employees as if they are responsible and reliable, they will be. He even wrote a book about it, subtitled: *L'entreprise qui croit que l'homme est bon*. Translation: 'The company that believes people are good.'

5

Companies like Buurtzorg and FAVI are proof that everything changes when you exchange suspicion for a more positive view of human nature.

Skill and competence become the leading values, not revenue or productivity. Just imagine what this would mean in other jobs and professions. CEOs would take the helm out of faith in their companies, academics would burn the midnight oil out of a thirst for knowledge, teachers would teach because they feel responsible for their students, psychologists would treat only as long as their patients require and bankers would derive satisfaction from the services they render.

Of course, there are already scores of teachers and bankers, academics and managers who are passionately motivated to help others. Not, however, *because* of the labyrinths of targets, rules, and procedures, but *despite* them.

Edward Deci, the American psychologist who flipped the script on how we think about motivation, thought the question should no longer be how to motivate others, but how we shape a society so that people motivate themselves. This question is neither conservative nor progressive, neither capitalist nor communist. It speaks to a new movement – a new realism. Because nothing is more powerful than people who do something *because they want to do it*.

14

Homo ludens

I

For days after my conversation with Jos de Blok, my mind kept coming back to the same question: what if the whole of society was based on trust?

To pull off a U-turn of this magnitude, we'd have to start at the beginning, I thought. We'd have to start with kids. But when I dived into the educational literature, I soon came up against a few harsh facts. Over the past decades, the intrinsic motivation of children has been systematically stifled. Adults have been filling children's time with homework, athletics, music, drama, tutoring, exam practice – the list of activities seems endless. That means less time for that one other activity: play. And then I mean play in the broadest sense – the freedom to go wherever curiosity leads. To search and to discover, to experiment and to create. Not along any lines set out by parents or teachers, but just because. For the fun of it.

Everywhere you look, children's freedom is being limited.[1] In 1971, 80 per cent of British seven- and eight-year-olds still walked to school on their own. These days it's a mere 10 per cent. A recent poll among twelve thousand parents in ten countries revealed that prison inmates spend more time outdoors than most kids.[2] Researchers at the University of Michigan found the time kids spent at school increased by 18 per cent from 1981 to 1997. Time spent on homework went up 145 per cent.[3]

Sociologists and psychologists alike have expressed alarm at these developments. One long-term American study found a diminishing 'internal locus of control' among children, meaning they increasingly feel their lives are being determined by others. In the US, this shift has been so seismic that in 2002 the average child felt less 'in control' than 80 per cent of kids did in the 1960s.[4]

Though these figures are less dramatic in my own country, the trend is the same. In 2018, Dutch researchers found that three in ten kids play outside either once a week or not at all.[5] A large study carried out by the OECD (the global think tank) among school kids, meanwhile, revealed that those in Holland are the least motivated of all countries surveyed. Tests and report cards have so blunted their intrinsic motivation that their attention evaporates when faced with an ungraded assignment.[6]

And that's to say nothing of the biggest shift of all: parents spending much more time with their kids. Time to read. Time to help out with homework. Time to take them to sports practice. In the Netherlands, the amount of time spent on parenting these days is over 150 per cent more than in the 1980s.[7] In the US, working mothers spend more time with their kids today than stay-at-home mothers did in the 1970s.[8]

Why? What's behind this shift? It's not as if parents suddenly gained oceans of time. On the contrary, since the 1980s parents everywhere have been working harder. Maybe that's the key: our fixation on work at the expense of everything else. As education policymakers began pushing rankings and growth, parents and schools became consumed with testing and results.

Kids are now being categorised at an ever younger age. Between those with heaps of ability and promise, and those with less. Parents worry: is my daughter being challenged enough? Is my son keeping up with his peers? Will they be accepted into a university? A recent study among 10,000 American students revealed that 80 per cent think their parents are more concerned about good grades than qualities like compassion and kindness.[9]

At the same time there's a widespread sense that something valuable is slipping away. Such as spontaneity. And playfulness. As a parent, you're bombarded with tips on how to steel yourself and your child against the pressures to achieve. There's a whole genre on how to work less and be more mindful. But what if a little self-help doesn't cut it?

To better understand what's going on, we need to define what we mean by play. Play is not subject to fixed rules and regulations, but is open-ended and unfettered. It's not an Astroturfed field with parents shouting at the sidelines; it's kids frolicking outside without parental supervision, making up their own games as they go along.

When kids engage in this kind of play, they think for themselves. They take risks and colour outside the lines, and in the process train their minds and motivation. Unstructured play is also nature's remedy against boredom. These days we give kids all kinds of manufactured entertainment, from the LEGO® Star Wars Snowspeeder™, complete with detailed assembly instructions, to the Miele Kitchen Gourmet Deluxe with electronic cooking sounds.

The question is, if everything comes prefabricated, can we still cultivate our own curiosity and powers of imagination?[10] Boredom may be the wellspring of creativity. 'You can't teach creativity,' writes psychologist Peter Gray, 'all you can do is let it blossom.'[11]

Among biologists, there's consensus that the instinct to play is rooted deep in our nature. Almost all mammals play, and many other animals also can't resist. Ravens in Alaska love whizzing down snow-covered rooftops.[12] On a beach in Australia, crocodiles have been spied surfing the waves for kicks, and Canadian scientists have observed octopuses firing water jets at empty medicine bottles.[13]

On the face of it, play may seem like a pretty pointless use of time. But the fascinating thing is that it's the most intelligent animals that exhibit the most playful behaviour. In Chapter 3 we saw that domesticated animals play their whole lives. What's more, no other species enjoys a childhood as long as *Homo puppy*. Play gives meaning to life, wrote the Dutch historian Johan Huizinga back in 1938. He christened us *Homo ludens* – 'playing man'. Everything we call 'culture,' said Huizinga, originates in play.[14]

Anthropologists suspect that for most of human history children were permitted to play as much as they pleased. Considerable though the differences between individual hunter-gatherer cultures may be, the culture of play looks very similar across the board.[15] Most significant of all, say researchers, is the immense freedom afforded to youngsters. Since nomads rarely feel they can dictate child development, kids are allowed to play all day long, from early in the morning until late into the night.

But are children equipped for life as an adult if they never go to school? The answer is that, in these societies, playing and learning are one and the same. Toddlers don't need tests or grades to learn to walk or talk. It comes to them naturally, because they're keen to explore the world. Likewise, hunter-gatherer children learn through play. Catching insects, making bows and arrows, imitating animal calls – there's so much to do in the jungle. And survival requires tremendous knowledge of plants and animals.

Equally, by playing together children learn to cooperate. Hunter-gatherer kids almost always play in mixed groups, with boys and girls and all ages together. Little kids learn from the big kids, who feel a responsibility to pass on what they know. Not surprisingly, competitive games are virtually unheard of in these societies.[16] Unlike adult tournaments, unstructured play continually requires participants to make compromises. And if anyone's unhappy, they can always stop (but then the fun ends for everyone).

2

The culture of play underwent a radical change when humans started settling down in one place.

For children, the dawn of civilisation brought the yoke of mind-numbing farm labour, as well as the idea that children required raising, much like one might raise tomatoes. Because if children were born wicked, then you couldn't leave them to their own devices. They first needed to acquire that veneer of civilisation, and often this called for a firm hand. The notion that parents should ever strike a child originated only recently, among our agrarian and city-dwelling ancestors.[17]

With the emergence of the first cities and states came the first education systems. The Church needed pious followers, the army loyal soldiers and the government hard workers. Play was the enemy, all three agreed. 'Neither do we allow any time for play,' dictated the English cleric John Wesley (1703–91) in the rules he established for his schools. 'He that plays when he is a boy, will play when he is a man.'[18]

Not until the nineteenth century was religious education supplanted by state systems in which, in the words of one historian, 'a French minister of education could boast that, as it was 10:20 a.m., he knew exactly which passage of Cicero all students of a certain form throughout France would be studying'.[19] Good citizenship had to be drilled into people from an early age, and those citizens also had to learn to love their country. France, Italy and Germany had all been traced out on the map; now it was time to forge Frenchmen, Italians and Germans.[20]

During the Industrial Revolution, a large portion of manufacturing drudgery was relegated to machines. (Not everywhere, of course – in Bangladesh kids still work sewing machines to produce our bargains.) That changed the aim of education. Now, children had to learn to read and write, to design and organise so that they could pay their own way when they were adults.

Not until the late nineteenth century did children once again have more time to play. Historians call this period the 'golden age' of unstructured play, when child labour was banned and parents increasingly left kids to themselves.[21] In many neighbourhoods in Europe and North America no one even bothered to keep an eye on them, and kids simply roamed free most of the day.

These golden days were short-lived, however, as from the 1980s onward life grew progressively busier, in the workplace and the classroom. Individualism and the culture of achievement gained precedence. Families grew smaller and parents began to worry whether their progeny would make the grade.

Kids who were too playful now might even be sent to a doctor. In recent decades, diagnoses of behavioural disorders have increased exponentially, of which perhaps the best example is ADHD. This is the only disorder, I once heard a psychiatrist remark, that's seasonal: what seems insignificant over summer vacation requires more than a few kids to be dosed on Ritalin when schools start again.[22]

Granted, we're a lot less strict with kids today than we were a hundred years ago, and schools are no longer the prisons they resembled in the nineteenth century. Kids who behave badly don't get a slap, but a pill. Schools no longer indoctrinate, but teach a more diverse curriculum than ever, transferring as much knowledge to students as possible so they'll find well-paying jobs in the 'knowledge economy'.

Education has become something to be endured. A new generation is coming up that's internalising the rules of our achievement-based society. It's a generation that's learning to run a rat race where the main metrics of success are your résumé and your pay cheque. A generation less inclined to colour outside the lines, less inclined to dream or to dare, to fantasise or explore. A generation, in short, that's forgetting how to play.

3

Is there another way?

Could we go back to a society with more room for freedom and creativity?

Could we build playgrounds and design schools that don't constrain but, rather, unchain our need to play?

The answer is yes, and yes, and yes.

Carl Theodor Sørensen, a Danish landscape architect, had already designed quite a few playgrounds before realising they bored kids senseless. Sandboxes, slides, swings ... the average playground is a bureaucrat's dream and a child's nightmare. Little wonder, Sørensen thought, that kids prefer to play in junkyards and building sites.

This inspired him to design something completely new at the time: a playground without rules or safety regulations. A place where kids themselves are in charge.

In 1943, in the midst of the German occupation, Sørensen tested his idea in the Copenhagen suburb of Emdrup. He filled a 75,000-square-foot lot with broken-down cars, firewood and old tyres. Children could smash, bang and tinker with hammers, chisels and screwdrivers. They could climb trees and build fires, dig pits and make huts. Or, as Sørensen later put it: they could 'dream and imagine and make dreams and imagination reality'.[23]

His 'junk playground' was a resounding success, on an average day attracting two hundred kids to Emdrup. And even with quite a few 'troublemakers', almost immediately it became apparent that 'the noise, screams and fights found in dull playgrounds are absent, for the opportunities are so rich that the children do not need to fight'.[24] A 'play leader' was employed to keep an eye on things, but he kept himself to himself. 'I cannot, and indeed will not, teach the children anything,' pledged the first play leader, John Bertelsen.[25]

Several months after the war ended, a British landscape architect paid a visit to Emdrup: Lady Allen of Hurtwood admitted she was 'completely swept off my feet' by what she encountered there.[26] In the years that followed, she wielded her influence to spread the gospel of junk, chanting, 'Better a broken bone than a broken spirit.'[27]

Soon, bombsites were being opened up to children all across Britain, from London to Liverpool, from Coventry to Leeds. Joyful shouts now sounded in places that only recently had reverberated with the death and destruction of German bombers. The new playgrounds became a metaphor for Britain's reconstruction and a testament to its resilience.

True, not everybody was enthusiastic. Adults always have two objections to these kinds of playgrounds. One: they're ugly. In fact, they're an eyesore. But where parents see disorder, kids see possibilities. Where adults can't stand filth, kids can't stand to be bored.

Objection two: junk playgrounds are dangerous. Protective parents feared that Emdrup would lead to a procession of broken bones and bashed brains. But after a year, the worst injuries required nothing more serious than a sticking plaster. One British insurance company was so impressed that it began charging junk playgrounds lower premiums than standard ones.[28]

Even so, by the 1980s, what in Britain became known as adventure playgrounds began to struggle. As safety regulations proliferated, manufacturers realised they could make a killing marketing self-styled 'safe' equipment. The consequence? These days there are considerably fewer Emdrups than there were forty years ago.

More recently, however, interest in Carl Theodor Sørensen's old idea has revived. And rightly so. Science has now supplied a mountain of evidence that unstructured, risky play is good for children's physical and mental wellbeing.[29] 'Of all the things I have helped to realise,' Sørensen concluded late in life, 'the junk playground is the ugliest, yet for me it is the best and most beautiful.'[30]

4

Could we take this a step further?

If kids can handle greater freedom outdoors, what about indoors? Many schools are still run like glorified factories, organised around bells, timetables and tests. But if children learn through play, why not model education to match? This was the question that occurred to Sjef Drummen, artist and school director, a few years ago.

Drummen is one of those people who never lost his knack for play, and who has always had an aversion to rules and authority. When he comes to pick me up from the railway station, he leaves his car parked flagrantly across the bike path. With me as his captive audience, he launches into a monologue that doesn't let up for the next few hours. Now and then I manage to get in a question. Grinning, he admits he's notorious for pushing his point.

But it wasn't Drummen's gift of the gab that got me on a train to the city of Roermond, in the southern reaches of the Netherlands. I came because something extraordinary is happening here.

Try to picture a school with no classes or classrooms. No homework or grades. No hierarchy of vice-principals and team leaders – only teams of autonomous teachers (or 'coaches' as they're called here). Actually, the students are the ones in charge. At this school, the director is routinely booted out of his office because the kids need it for a meeting.

And, no, this isn't one of those elite private schools for off-beat students with zany parents. This school enrols kids from all backgrounds. Its name? Agora.

It all started in 2014, when the school decided to tear down the dividing walls. (Drummen: 'Shut kids up in cages and they behave like rats.') Next, kids from all levels were thrown together. ('Because that's what the real world is like.') Then each student had to draw up an individual plan. ('If your school has one thousand kids, then you have one thousand learning pathways.')

The result?

Upon entering the school, what first comes to mind is a junk playground. Rather than rows of seats lined up facing the board, I see a colourful chaos of improvised desks, an aquarium, a replica of Tutankhamun's tomb, Greek columns, a bunk bed, a Chinese dragon and the front half of a sky-blue '69 Cadillac.

One of the students at Agora is Brent. Now seventeen, until a few years ago he attended a bilingual college prep school where he was earning good grades in everything except French and German – which he was failing. Under the Dutch three-track system, Brent was transferred down to a general secondary education track and then, when he continued to lag behind, to a vocational track. 'When they told me, I ran home, furious. I told my mother I was getting a job at McDonald's.'

But thanks to friends of friends, Brent wound up at Agora, where he was free to learn what *he* wanted. Now he knows all about the atomic bomb, is drafting his first business plan and can carry on a conversation in German. He's also been accepted on an international programme at Mondragon University in Shanghai.

According to his coach Rob Houben, Brent felt conflicted about announcing his admission to college. 'He told me, "There's still so much I want to give back to this school for everything it's done for me".'

Or take Angelique, aged fourteen. Her primary school sent her to vocational education, but the girl I meet is terrifically ana-lytical. She's obsessed with Korea for some reason and set on studying there, and has already taught herself quite a bit of the language. Angelique is also vegan and has compiled an entire book of arguments to fire at meat-eaters. (Coach Rob: 'I always lose those debates.')

Every student has a story. Rafael, also fourteen, loves pro-gramming. He shows me a security leak he discovered on the Dutch Open University website. He notified the webmaster, but it hasn't been fixed yet. Laughing, Rafael tells me, 'If I wanted to get his attention, I could change his personal password.'

When he shows me the website of a company he's done some front-end work on, I ask if he shouldn't be billing them for his trouble. Rafael gives me an odd look. 'What, and lose my motivation?'

More than their sense of purpose, what impresses me most about these kids is their sense of community.

Several of the students I talk to would probably have been picked on mercilessly at my old school. But at Agora no one gets bullied, everybody I talk to says so. 'We set each other straight,' says Milou, aged fourteen.

Bullying is often regarded as a quirk of our nature; something that's part and parcel of being a kid. Not so, say sociologists, who over the years have compiled extensive research on the places where bullying is endemic. They call these *total institutions*.[31] Sociologist Erving Goffman, writing some fifty years ago, described them as follows:

- Everybody lives in the same place and is subject to a single authority.
- All activities are carried out together and everybody does the same tasks.
- Activities are rigidly scheduled, often from one hour to the next.
- There is a system of explicit, formal rules imposed by an authority.

Of course, the ultimate example is a prison, where bullying runs rampant. But total institutions show up in other places, too, such as nursing homes. The elderly, when penned together, can develop caste systems in which the biggest bullies claim the best seats and tables at Bingo time.[32] One American expert on bullying even calls Bingo 'the devil's game'.[33]

And then there are schools. Bullying is by far the most pervasive at typical British boarding schools (the kind that inspired William Golding's *Lord of the Flies*).[34] And little wonder: these schools resemble nothing so much as prisons. You can't leave, you have to earn a place in a rigid hierarchy, and there's a strict division between pupils and staff. These competitive institutions are part and parcel of Britain's upper-class establishment – many London politicians went to boarding school – but according to education scientists, they thwart our playful nature.[35]

The good news is things can be different. Bullying is practically non-existent at unstructured schools like Agora. Here, you can take a breather whenever you need one: the doors are always open. And, more importantly, everyone here is different. Difference is normal, because children of all ages, abilities and levels intermingle.

'At my old school,' says Brent, 'you didn't talk to kids in vocational ed.' Then he and Joep (fifteen) tell me about the time Noah (fifteen, originally placed in a vocational programme) gave them a lecture on a skill they sorely lack: planning. 'Noah had planned out the whole next year and a half of his life,' explains Joep. 'We learned a lot.'

The longer I walk around Agora, the more it hits me how crazy it is to corral kids by age and ability. Experts have for years been warning about the growing gap between well-educated and less educated segments of the population, but where does this rift actually begin? Jolie (fourteen) says, 'I don't see the difference. I've heard vocational students say things that make much more sense than the so-called honours kids.'

Or take the customary way schools chop days up into timed periods. 'Only at school is the world divided up into subject chunks,' notes Coach Rob. 'Nowhere else does that happen.' At most schools, just when a student finds their flow, the bell rings for the next class. Could there be a system more rigged to discourage learning?

Before you get the wrong idea, it's important not to exaggerate Agora's laissez-faire philosophy. The school may promote freedom, but it's not a free for all. There's a minimal yet vital structure. Every morning, a student opens the school day. There's one hour of quiet time daily, and every student meets with their coach once a week. Moreover, expectations are high, the kids know it, and they work with the coaches to set personal goals.

These coaches are essential. They nurture and challenge, encourage and guide. In all honesty, their job looks harder than ordinary teaching. For starters, they have to unlearn much of their training as teachers. 'Most of what kids want to learn, you can't teach them,' explains Rob. He doesn't speak Korean, for example, and knows nothing about computer programming, but nonetheless he's helped Angelique and Rafael on their respective paths.

The big question, of course, is: would this model work for most children?

Given the incredible diversity in the student body at Agora, I see every reason to believe it might.[36] The kids say it took some to get used to, but they've learned to follow where their curiosity leads. Sjef Drummen compares it to caged chickens at battery farms: 'A couple years ago I bought some off a farmer. When I let them out in my yard, they just stood there for hours, nailed to the spot. It took a week before they found the courage to move.'

And now the bad news. Any kind of radical renewal inevitably clashes with the old system.

In truth, Agora is educating kids for a very different kind of society. The school wants to give them room to become autonomous, creative, engaged citizens. But if Agora doesn't meet standardised testing criteria, the school won't pass inspection and can wave its government funding goodbye. This is the mechanism that's consistently put the brakes on initiatives like Agora.

So maybe there's an even bigger question we should be asking: What's the purpose of education? Is it possible we've become transfixed on good grades and good salaries?

In 2018 two Dutch economists analysed a poll of twenty-seven thousand workers in thirty-seven countries. They found that fully a quarter of respondents doubt the importance of their own work.[37] Who are these people? Well, they're certainly not cleaners, nurses, or police officers. The data show that most 'meaningless jobs' are concentrated in the private sector – in places like banks, law firms and ad agencies. Judged by the criteria of our 'knowledge economy', the people holding these jobs are the definition of success. They earned straight As, have sharp LinkedIn profiles and take home fat pay cheques. And yet the work they do is, by their own estimation, useless to society.

Has the world gone nuts? We spend billions helping our biggest talents scale the career ladder, but once at the top they ask themselves what it's all for. Meanwhile, politicians preach the need to secure a higher spot in international country rankings, telling us we need to be more educated, earn more money and bring the economy more 'growth'.[38]

But what do all those degrees really represent? Are they proof of creativity and imagination, or of an ability to sit still and nod? It's like the philosopher Ivan Illich said decades ago: 'School is the advertising agency which makes you believe that you need the society as it is.'[39]

Agora, the playing school, proves there is a different way. It's part of a movement of schools that are charting an alternative course. People may scoff at their approach to education, but there's plenty of evidence it works: Summerhill School in Suffolk, England, has been demonstrating since 1921 that kids can be entrusted with an abundance of freedom. And so has the Sudbury Valley School in Massachusetts, where since the 1960s thousands of kids have spent their youth – and gone on to lead fulfilling lives.[40]

The question is not: can our kids handle the freedom?

The question is: do we have the courage to give it to them?

It's an urgent question. 'The opposite of play is not work,' the psychologist Brian Sutton-Smith once said. 'The opposite

of play is depression.'[41] These days, the way many of us work –
with no freedom, no play, no intrinsic motivation – is fuelling
an epidemic of depression. According to the World Health
Organization, depression is now the number one global disease.[42] Our biggest shortfall isn't in a bank account or budget
sheet, but inside ourselves. It's a shortage of what makes life
meaningful. A shortage of play.

Visiting Agora has made me see there's a ray of hope. Later,
when Sjef Drummen drops me off back at the station, he gives
me another grin. 'I think I talked your ear off today.' True,
but I have to hand it to him: walk around his school for any
length of time and you'll feel quite a few old certainties start to
crumble.

But now I understand: this is a journey back to the beginning. Agora has the same teaching philosophy as hunter-
gatherer societies. Children learn best when left to their own
devices, in a community bringing together all ages and abilities
and supported by coaches and play leaders.[43] Drummen calls it
'Education 0.0' – a return to *Homo ludens*.

15

This Is What Democracy Looks Like

I

It was an unlikely setting for a revolution. The municipality in western Venezuela had a population of less than two hundred thousand, and a small elite had been calling the shots for hundreds of years.[1] Yet it was here in Torres that ordinary citizens found an answer to some of the most urgent questions of our times.

How can we restore trust in politics? How can we stem the tide of cynicism in society? And how can we save our democracy?

Democracies around the globe are afflicted by at least seven plagues. Parties eroding. Citizens who no longer trust one another. Minorities being excluded. Voters losing interest. Politicians who turn out to be corrupt. The rich getting out of paying taxes. And the growing realisation that our modern democracy is steeped in inequality.

Torres found a remedy for all these problems. Tried and tested now for twenty-five years, the Torres solution is mind-bogglingly simple. It's being adopted around the world, yet rarely makes the news. Perhaps because, like Buurtzorg and Agora, it's a realistic initiative premised on a fundamentally different view of human nature. One that doesn't see people as complacent or reduce them to angry voters, but instead asks, *what if there's a constructive and conscientious citizen inside each of us?*

Put differently: what if real democracy's possible?

The story of Torres began on 31 October 2004. Election day. Two opposing candidates were running for mayor of the Venezuelan municipality: The incumbent Javier Oropeza, a wealthy landowner backed by the commercial media, and Walter Cattivelli, who was endorsed by reigning president Hugo Chávez's powerful party.

It wasn't much of a choice. Oropeza or Cattivelli – either way, the corrupt establishment would continue to run the show. Certainly there was nothing to hint that Torres was about to reinvent the future of democracy.

Actually, there was another candidate, albeit one hardly worth mentioning. Julio Chávez (no relation) was a marginal agitator whose supporters consisted of a handful of students, cooperatives and union activists. His platform, which could be summed up in a single sentence, was downright laughable. If he was elected mayor, Julio would hand over power to the citizens of Torres.

His opponents didn't bother to take him seriously. Nobody thought he stood a chance. But sometimes the biggest revolutions begin where you least expect them. That Sunday in October, with just 35.6 per cent of the vote in this three-way race, Julio Chávez was narrowly elected mayor of Torres.[2]

And he kept his word.

The local revolution began with hundreds of gatherings. All residents were welcome – not only to debate issues, but to make real decisions. One hundred per cent of the municipal investment budget, roughly seven million dollars, was theirs to spend.

It was time, announced the new mayor, for a *true* democracy. Time for airless meeting rooms, lukewarm coffee, fluorescent lighting and endless bookkeeping. Time for government not by public servants and career politicians, but by the citizens of Torres.

The old elite looked on in horror as their corrupt system was taken apart. '[They] said that this was anarchy,' recalled Julio (everybody calls this mayor by his first name) in an interview with an American sociologist. 'They said that I was crazy to give up my power.'[3]

The governor of the state of Lara, of which Torres was a part, was furious that Oropeza, his puppet, had been bested by this upstart. He decided to cut off the municipality's funding and appoint a new council. But he hadn't reckoned on the groundswell of support for the freshly elected mayor. Hundreds of residents marched on city hall, refusing to go home until their budget was adopted.

In the end, the people won. Within ten years of Julio Chávez's election, Torres had pulled off several decades' worth of progress. Corruption and clientelism were way down, demonstrated a University of California study, and the population was participating in politics like never before. New houses and schools were going up, new roads were being built and old districts were getting spruced up.[4]

To this day Torres has one of the largest participatory budgets in the world. Some fifteen thousand people provide input, and assemblies are held early each year in 560 locations across the municipality. Everyone is welcome to submit proposals and elect representatives. Together, the people of Torres decide where to allocate their millions in tax revenue.

'In the past, government officials would stay in their air-conditioned offices all day and make decisions there,' one resident said. 'They never even set foot in our communities. So who do you think can make a better decision about what we need, an official in his office who has never come to our community, or someone who is from the community?'[5]

2

Now, you may be thinking to yourself: nice anecdote, but one swallow doesn't make a democratic summer. So some obscure place ventured off the beaten track, why is *that* a revolution?

The thing is, what happened in Torres is just one instance among many. The bigger story started years earlier, when a metropolis in Brazil took the unprecedented step of entrusting a quarter of its budget to the populace. That city was Porto Alegre and the year was 1989. A decade later, the idea had been copied by more than a hundred cities across Brazil, and from there it began to spread around the world. By 2016, more than fifteen hundred cities, from New York City to Seville and from Hamburg to Mexico City, had enacted some form of participatory budgeting.[6]

What we're talking about here is in fact one of the biggest movements of the twenty-first century — but the chances are you've never heard of it. It's just not juicy enough for the news. Citizen politicians don't have reality-star appeal or money for spin doctors and ad campaigns. They don't devise pithy one-liners to throw around in so-called debates, and they couldn't care less about daily polls.

What citizen politicians do is engage in calm and deliberative dialogue. This may sound dull, but it's magic. It might just be the remedy for the seven plagues afflicting our tired old democracies.

1. *From cynicism to engagement*

In most countries there's a deep divide between the people and the political establishment. With the suits in Washington, Beijing and Brussels making most of the decisions, is it any wonder the average person feels unheard and unrepresented?

In Torres and Porto Alegre, almost everyone is personally acquainted with a politician. Since some 20 per cent of the population has participated in city budgeting, there's also less grumbling about what politicians are doing wrong.[7] Not happy about how things are going? Help fix it. 'It's not the suits who come here and tell us what to do. It's us,' reported a participant in Porto Alegre. 'I am a humble person. I have participated since the beginning. [...] [Budgeting] makes people talk, even the poorest.'[8]

At the same time, trust in the city council has gone up in Porto Alegre. And mayors are among those to profit most, a Yale political scientist discovered, because mayors who empower their citizens are more likely to be re-elected.[9]

2. *From polarisation to trust*

When Porto Alegre launched its participatory budgeting experiment, the city was not exactly a bastion of trust. In fact, there are few countries where people trust each other less than in Brazil.[10] Most experts therefore rated the city's chances of pulling off a democratic spring as slim to none. People first had to band together, form clubs, tackle discrimination and so on. That would then prepare the ground for democracy to take root.[11]

Porto Alegre turned this equation around. Only *after* the administration launched a participatory budget did trust begin to grow. Community groups then multiplied, from 180 in 1986 to 600 in 2000. Soon, engaged citizens were addressing each other as *companheiro* – as compatriots and brothers.

The people of Porto Alegre behaved much like Agora founder Sjef Drummen's formerly caged chickens. When first released from their coops, they stood nailed to the ground. But they soon found their feet. 'The most important thing,' said one, 'is that more and more people come. Those who come for the first time are welcome. You have the responsibility of not abandoning [them]. That's the most important thing.'[12]

3. *From exclusion to inclusion*

Political debates can be so complex that people have a hard time following along. And in a diploma democracy, those with little money or education tend to be sidelined. Many citizens of democracies are, at best, permitted to choose their own aristocracy.

But in the hundreds of participatory budgeting experiments, it's precisely the traditionally disenfranchised groups that are well represented. Since its 2011 start in New York City, the meetings have attracted chiefly Latinos and African Americans.[13] And in Porto Alegre, 30 per cent of participants come from the poorest 20 per cent of the population.[14]

'The first time I participated I was unsure,' admitted one Porto Alegre participant, 'because there were persons there with college degrees, and we don't have [degrees]. [...] But with time, we started to learn.'[15] Unlike the old political system, the new democracy is not reserved for well-off white men. Instead, minorities and poorer and less-educated segments of society are far better represented.

4. *From complacency to citizenship*

On the whole, voters tend to take a fairly dim view of politicians, and vice versa. But democracy as practised in Torres and Porto Alegre is a training ground for citizenship. Give people a voice in how things are run and they become more nuanced about politics. More sympathetic. Even smarter.

A journalist reporting on participatory budgeting in Vallejo, California, expressed astonishment at the level of people's commitment: 'here were all these people of different ages and ethnic groups, who could be home watching their local baseball team in the World Series, who were instead talking about rules and voting procedure. And not only that, they were passionate about it.'[16]

Time and again, researchers remark on the fact that almost everybody has something worthwhile to contribute – regardless of formal education – as long as everyone's taken seriously.

5. From corruption to transparency

Before participatory budgeting came to Porto Alegre, citizens who wanted a politician's ear could expect to spend hours waiting outside their office. And then it helped to have a wad of cash to pass under the table.

According to a Brazilian sociologist who spent years researching Porto Alegre, the participatory process undermined the old culture of greasing palms. People were better informed about civic finances, and that made it harder for politicians to accept bribes and award jobs.[17]

'We see it [the participatory budget] as an organizing tool,' said a Chicago resident. 'It will help our members learn more about the city budget and then we can press the alderman about other things he controls.'[18] In other words: participatory budgeting bridges the divide between politics and the people.

6. From self-interest to solidarity

How many piles of books have been written in recent years about the fragmentation of society? We want better healthcare, better education and less poverty, but we also have to be willing to pitch in.

Unbelievable as it may sound, studies find that participatory budgeting actually makes people more willing to pay taxes. In Porto Alegre, citizens even asked for *higher* taxes – something political scientists had always deemed unthinkable.[19]

'I had not understood that council tax paid for so much,' enthused one participant in Leicester East (UK). 'It was very good to find out the services it pays for.'[20] This redefines taxes into a contribution you pay as a member of society. Many of those involved in participatory budgeting say the experience made them feel like real citizens for the first time. After a year, as one Porto Alegren put it, you learn to look beyond your own community: 'You have to look at the city as a whole.'[21]

7. From inequality to dignity

Before Porto Alegre embarked on its democratic adventure, the city was in dire financial straits. One-third of the population was living in slums.

But then things began changing fast – much faster than in cities that didn't adopt a participatory budget.[22] Access to running water went from 75 per cent in 1989 to 95 per cent in 1996, and access to city sewage service went from a measly 48 per cent to 95 per cent of the population. The number of children attending school tripled, the number of roads built multiplied fivefold, and tax evasion plummeted.[23]

Thanks to citizen budgets, less public money went into prestige projects like real estate. The World Bank found that more went to infrastructure, education and healthcare, particularly in poorer communities.[24]

In 2014 an American research team published the first large-scale study on the social and economic impact of participatory budgeting across Brazil. Their conclusion was loud and clear: 'We find PB programs are strongly associated with increases in health care spending, increases in civil society organizations, and decreases in infant mortality rates. This connection strengthens dramatically as PB programs remain in place over longer time frames.'[25]

In the mid-1990s, Britain's Channel 4 launched a new TV programme called *The People's Parliament*. The show randomly invited hundreds of Britons from all walks of life to go head-to-head on controversial issues like drugs, arms sales and juvenile crime. At the end of each episode, they had to reach a compromise.

According to the *Economist*, 'Many viewers of the *People's Parliament* have judged its debates to be of higher quality than those in the House of Commons. Members of the former, unlike the latter, appear to listen to what their fellows say.'[26] So what did Channel 4 do? It pulled the plug. Producers felt the debates were too calm, too thoughtful, too sensible, and much preferred the kind of confrontational entertainment we call 'politics'.

But participatory democracy isn't an experiment cooked up for TV. It's a sound method for tackling the plagues of the old democracy.

Like any other, this form of democracy has its shortcomings. The focus on yearly investments can come at the cost of a city's long-term vision. More importantly, many participatory processes are too limited. Porto Alegre's budget was curtailed in 2004 when a conservative coalition came to power, and now it's unclear if the tradition will survive in the city where it all started.

Sometimes participatory budgeting is used as a cover-up – a sham concession by elites who, behind the scenes, are still running the show. Then citizens' assemblies only serve to rubber-stamp decisions already taken. Naturally, this engenders cynicism, but it doesn't legitimate denying citizens a direct voice. 'Treat responsible citizens as ballot fodder and they'll behave like ballot fodder,' writes historian David Van Reybrouck, 'but treat them as adults and they'll behave like adults.'[27]

3

It was back in fourth grade that Mr Arnold taught us about communism. 'From each according to his ability, to each according to his needs.' Or as I read (years later) in the *Oxford English Dictionary*: 'A theory or system of social organization in which all property is owned by the community and each person contributes and receives according to their ability and needs.'[28]

As a child, this sounded like a great idea. Why not share everything? But in the years that followed, like so many kids I had to face a disappointing realisation: sharing everything equally may be a fine idea, in practice it degenerates into chaos, poverty, or worse – a bloodbath. Look at Russia under Lenin and Stalin. China under Mao. Cambodia under Pol Pot.

These days, the C-word tops the list of controversial ideologies. Communism, we're told, *cannot* work. Why? Because it's based on a flawed understanding of human nature. Without private property, we lose all motivation and swiftly revert into apathetic parasites.

Or so the story goes.

Even as a teenager it struck me as odd that the case for communism's 'failure' seemed to rest solely on the evidence of bloodthirsty regimes in countries where ordinary citizens had no say – regimes supported by all-powerful police states and corrupt elites.

What I didn't realise back then was that communism – according to the official definition, at least – has been a successful system for hundreds of years, one that bears little resemblance to the Soviet Union. In fact, we practise it every day. Even after decades of privatisation, big slices of our economy still operate according to the communist model. This is so normal, so obvious, that we no longer see it.

Simple example: you're sitting at the dinner table and can't reach the salt. You say, 'Please pass the salt' and, just like that, someone hands you the salt – for free. Humans are crazy about this kind of *everyday communism*, as anthropologists call it, sharing our parks and plazas, our music and stories, our beaches and beds.[29]

Perhaps the best example of this liberality is the household. Billions of homes worldwide are organised around the communist principle: parents share their possessions with their children and contribute as they're able. This is where we get the word 'economy', which derives from the Greek *oikonomíā*, meaning 'management of a household'.

In the workplace we're also constantly showing our communist colours. While writing this book, for instance, I benefited from the critical eyes of dozens of colleagues, who didn't ask a penny for their time. Businesses, too, are big fans of internal communism, simply because it's so efficient.

But what about strangers? After all, we don't share everything with everyone. On the other hand, how often have you charged people who asked for directions? Or when you held the door open for someone, or allowed another person to shelter under your umbrella? These are not tit-for-tat transactions; you do them because they're the decent thing to do, and because you believe other strangers would do the same for you.

Our lives are filled with these kinds of communist acts. The word 'communism' comes from the Latin *communis*, meaning 'communal'. You could see communism as the bedrock on which everything else – markets, states, bureaucracy – is built. This may help explain the explosion of cooperation and altruism that happen in the wake of natural disasters, such as in New Orleans in 2005. In a catastrophe, we go back to our roots.

Of course, we can't always apply the communist ideal of 'from each according to his ability, to each according to his needs', just as not everything can be assigned a monetary value. Zoom out, however, and you'll realise that on a day-to-day basis we share more with one another than we keep for ourselves.

This communal basis is a vital mainstay of capitalism. Consider how many companies are utterly dependent on the generosity of their customers. Facebook would be worth far less without the pictures and videos that millions of users share for free. And Airbnb wouldn't survive long without the innumerable reviews travellers post for nothing.

So why are we so blind to our own communism? Maybe it's because the things we share don't seem all that remarkable. We take sharing them for granted. Nobody has to print flyers explaining to people that it's nice to take a stroll in Central Park. Clean air has no need for public service announcements instructing you to inhale it. Nor do you think of that air – or the beach you relax on or the fairy tales you recount – as belonging to somebody.

It's only when someone decides to rent out the air, appropriate the beach, or claim the rights to the fairy tale that you take notice. Wait a minute, you think, didn't this belong to all of us?

The things we share are known as *the commons*. They can include just about anything – from a community garden to a website, from a language to a lake – as long as it's shared and democratically managed by a community. Some commons are part of nature's bounty (like drinking water), others are human inventions (websites like Wikipedia).

For millennia, the commons constituted almost everything on earth. Our nomadic ancestors had scarcely any notion of private property and certainly not of states. Hunter-gatherers viewed nature as a 'giving place' that provided for everybody's needs, and it never occurred to them to patent an invention or a tune. As we saw in Chapter 3, *Homo puppy* owes its success to the fact that we're master plagiarists.

It's only in the past 10,000 years that steadily bigger slices of the commons have been swallowed up by the market and the state. It began with the first chieftains and kings, who laid claim to lands which had previously been shared by everyone. Today, it's mainly multinationals that appropriate all kinds of commons, from water sources to lifesaving drugs and from new scientific knowledge to songs we all sing. (Like the nineteenth-century hit 'Happy Birthday', to which, up until 2015, the Warner Music Group owned the rights, thus raking in tens of millions in royalties.)

Or take the rise of the advertising industry, which has plastered unsightly billboards all over cities across the world. If someone sprays your house with graffiti, we call it vandalism. But for advertising you're allowed to deface public space and economists will call it 'growth'.

The concept of the commons gained currency with a piece published in the journal *Science* by American biologist Garrett Hardin. This was 1968, a time of revolution. Millions of demonstrators around the world took to the streets in protest, rallying to the cry: 'Be realistic. Demand the impossible.'

But not the conservative Garrett Hardin. His six-page paper made short work of hippie idealism. Title? 'The Tragedy of the Commons'.

'Picture a pasture open to all,' Hardin wrote. 'It is to be expected that each herdsman will try to keep as many cattle as possible on the commons.' But what makes sense at the individual level results in a collective disaster, with overgrazing leaving nothing but barren wasteland. Hardin used the term 'tragedy' in the Greek sense, to mean a regrettable but inevitable event: 'Freedom in a commons,' he said, 'brings ruin to all.'[30]

Hardin was not afraid to reach harsh conclusions. To the question should countries send food aid to Ethiopia, his response was: don't even start. More food would mean more children would mean more famine.[31] Like the Easter Island pessimists, he saw overpopulation as the ultimate tragedy, and restriction of reproductive rights as the solution. (Though not for him-self: Hardin fathered four children.)

It's hard to overstate the impact of Hardin's paper, which went on to become the most widely reprinted ever published in a scientific journal, read by millions of people across the world.[32] '[It] should be required reading for all students,' declared an American biologist in the 1980s, 'and, if I had my way, for all human beings.'[33]

Ultimately, 'The Tragedy of the Commons' would prove among the most powerful endorsements for the growth of the market and the state. Since common property was tragically doomed to fail, we needed either the visible hand of the state to do its salutary work, or the invisible hand of the market to save us. It seemed these two flavours – the Kremlin or Wall Street – were the only options available. Then, after the Berlin Wall came down in 1989, only one remained. Capitalism had won, and we became *Homo economicus*.

4

To be fair, at least one person was never swayed by Garrett Hardin's arguments.

Elinor Ostrom was an ambitious political economist and researcher at a time when universities didn't exactly welcome women. And, unlike Hardin, Ostrom had little interest in theor-etical models. She wanted to see how real people behave in the real world.

It didn't take her long to realise there was one crucial detail Hardin's paper had overlooked. Humans can talk. Farmers and fishermen and neighbours are perfectly capable of making agreements to keep their fields from turning into deserts, their lakes from being overfished and their wells from drying up. Just as the Easter Islanders continued to pull together, and participatory budgeters make decisions through constructive dialogue, so ordinary people successfully manage all manner of commons.

Ostrom set up a database to record examples of commons from all over the world, from shared pastures in Switzerland and cropland in Japan to communal irrigation in the Philippines and water reserves in Nepal. Everywhere she looked, Ostrom saw that pooling resources is by no means a recipe for tragedy, as Hardin contended.[34]

Sure, a commons can fall victim to conflicting interests or greed, but that's far from inevitable. All told, Ostrom and her team compiled more than five thousand examples of working commons. Many went back centuries, like the fishermen in Alanya, Turkey, who have a time-honoured tradition of drawing lots for fishing rights, or the farmers in the Swiss village of Törbel who jointly coordinate use of scarce firewood.

In her groundbreaking book *Governing the Commons* (1990), Ostrom formulated a set of 'design principles' for successful commons. A community must have a minimum level of autonomy, for instance, and an effective monitoring system. But she stressed that there's no blueprint for success, because the characteristics of a commons are ultimately shaped by the local context.

Over time, even Ostrom's department at the university began to resemble a commons. In 1973, she and her husband established what they called the Workshop in Political Theory and Policy Analysis at Indiana University, drawing academics from all over the world to study the commons. This workshop – a form chosen because the university had no rules dictating the structure of workshops – became a hive of discussion and discovery. In fact, it grew into something of an academic hippie commune, with parties where Ostrom herself led the singing of folksongs.[35]

And then one day, years later, the call came from Stockholm. Elinor Ostrom won the 2009 Nobel Prize in Economics, the first woman ever to win.[36] This choice sent a strong message. After the fall of the Berlin Wall in 1989 and the crash of capitalism in 2008, finally the moment had arrived to give the commons – that alternative between the state and the market – the spotlight it deserves.

5

It may not be breaking news, but since then the commons has made a spectacular comeback.

If it seems like history's repeating itself, that's because it's not the first time this has happened. In the late Middle Ages, Europe witnessed an explosion of communal spirit, in what historian Tine de Moor has called a 'silent revolution'. During this period, from the eleventh to the thirteenth centuries, an increasing share of pastureland came under collective control, and water boards, guilds and lay group homes sprouted up like mushrooms.[37] These commons worked well for hundreds of years, until they came under pressure in the eighteenth century.

Enlightenment-era economists decided collective farmlands were not maximising their production potential, so they advised governments to create *enclosures*. That meant cutting collective property into parcels to be divvied up among wealthy landowners, under whose stewardship productivity would grow.

Do you think capitalism's ascent in the eighteenth century was a natural development? Hardly. It wasn't the invisible hand of the market that gently shepherded peasants from their farms into factories, but the ruthless hand of the state, bayonet at the ready. Everywhere in the world, that 'free market' was planned and imposed from the top down.[38] It wasn't until the end of the nineteenth century that scores of unions and worker cooperatives began forming – spontaneously and from the bottom up – laying the basis for the twentieth century's system of social safety nets.

The same thing's happening again now. Following a period of enclosures and market forces (planned top-down by states), a silent revolution has been simmering at the bottom, giving rise in recent years, and particularly since the 2008 financial crisis, to an explosion of initiatives like care cooperatives, sick day pools and energy co-ops.

'History teaches us that man is essentially a cooperative being, a *homo cooperans*,' points out Tine de Moor. 'We have been building institutions that are focused on long-term cooperation for a long time now, in particular after periods of accelerated market development and privatisation.'[39]

So do we want less communism, or more?

In my high school economics class, we were taught that selfishness is in our nature. That capitalism is rooted in our deepest instincts. Buying, selling, doing deals – we're always out to maximise personal profit. Sure, we were told, the state can sprinkle a dusting of solidarity over our natural inclinations, but this can only happen from on high, and never without supervision and bureaucracy.

Now it turns out that this view is completely upside down. Our natural inclination is for solidarity, whereas the market is imposed from on high. Take the billions of dollars pumped in recent decades into frenzied efforts to turn healthcare into an artificial marketplace. Why? Because we have to be taught to be selfish.

That's not to say there aren't abundant examples of healthy and effective markets. And, of course, we must not forget that the rise of capitalism over the past two hundred years has brought huge gains in prosperity. De Moor therefore advocates what she calls 'institutional diversity', which recognises that while markets work best in some cases and state control is better in others, underpinning it all there has to be a strong communal foundation of citizens who decide to work together.

At this point, the future of the commons is still uncertain. Even as communal interests are making a comeback, they're also under siege. By multinationals, for instance, that are buying up water supplies and patenting genes, by governments that are privatising whatever they can get a buck for, and by universities that are selling off their knowledge to the highest bidder. Also by the advent of platform capitalism, which is enabling the likes of Airbnb and Facebook to skim the fat off the prosperity of the *Homo cooperans*. All too often, the sharing economy turns out to be more like a shearing economy – we all get fleeced.

For the moment, we're still locked in a fierce and undecided contest. On one side are the people who believe the whole world is destined to become one big commune. These are the optimists – also known as post-capitalists, presumably because communism is still a dirty word.[40] On the other side are the pessimists who foresee continued raids on the commons by Silicon Valley and Wall Street and ongoing growth in inequality.[41]

Which side will turn out to be right? Nobody really knows. But my money's on Elinor Ostrom, who was neither an optimist nor a pessimist, but a possibilist. She believed there was another way. Not because she subscribed to some abstract theory, but because she'd seen it with her own eyes.

6

One of the most promising alternatives to the existing capitalist model has actually been around for quite a while. You won't find it in progressive Scandinavia, in red China or in Latin America's cradles of anarchy – no, this alternative comes, rather unexpectedly, from a US state where terms like 'progressive' and 'socialist' are used as insults. From Alaska.

The idea started with Republican governor Jay Hammond (1922–2005), a hard-as-nails fur trapper and former fighter pilot who'd fought the Japanese in the Second World War. When in the late 1960s huge oil reserves were discovered in his home state, he decided this oil belonged to all Alaskans and proposed to put the profits into a great big communal piggy bank.

This piggy bank became the Alaska Permanent Fund, established in 1976. The next question, of course, was what to do with all the cash. Many conservative Alaskans opposed handing it over to the state, which would only fritter it away. But there was another option. Starting in 1982, every citizen of Alaska received an annual dividend in their bank account. In a good year, it could be as much as $3,000.

To this day, the Permanent Fund Dividend – PFD for short – is wholly unconditional. It's not a privilege, but a right. That makes the Alaskan model the polar opposite of the old-fashioned welfare state. Normally, you first have to prove you're sick enough, disabled enough, or otherwise needy enough to merit assistance, and not until you file dozens of forms testifying that you are past all hope do you get a scant amount of money.

That kind of system is primed to make people sad, listless and dependent, while an unconditional dividend does something else entirely. It fosters trust. Of course there were people who cynically assumed their fellow Alaskans would squander the dividend on alcohol and drugs. But as the realists observed, that's not what happened.

Most Alaskans invested their dividends in education and their children. In-depth analysis by two American economists showed that the PFD had no adverse effects on employment and that it substantially reduced poverty.[42] Research on comparable cash payments in North Carolina even revealed a slew of positive side effects. Healthcare costs went down, and kids performed better at school, effectively recouping the initial investment cost.[43]

What if we take Alaska's communal property philosophy and apply it more broadly? What if we say that groundwater, natural gas, the patents made possible by taxpayer money, and so much more, all belong to the community? Whenever a part of those commons is appropriated, or the planet polluted, or CO_2 dumped into our atmosphere, shouldn't we then – as members of the community – be compensated?[44]

A fund like this could have another, much bigger, payoff for all of us. This citizen's dividend, this unconditional payment premised on trust and belonging, would give each of us the freedom to make our own choices. Venture capital for the people.

In Alaska in any case, the PFD has clearly been a big hit. Any politician who even thinks of tampering with it risks career suicide.[45] Some will say that's because everyone's looking out for themselves. But maybe it's so popular because – like the real democracies of Porto Alegre and Torres – it goes beyond the old opposition between left and right, market and state, capitalism and communism. This is a different road, heading towards a new society, in which everyone has a share.

Part Five

THE OTHER CHEEK

'If you are to punish a man retributively you must injure him. If you are to reform him you must improve him. And men are not improved by injuries.'

George Bernard Shaw (1856–1950)

Not long ago, Julio Diaz, a young social worker, was taking the subway from work to his home in the Bronx in New York. As he did almost every day, he got off one stop early to grab a bite at his favourite diner.

But tonight wasn't like other nights. As he made his way to the restaurant from the deserted subway station, a figure jumped out from the shadows. A teenager, holding a knife. 'I just gave him my wallet,' Julio later told a journalist. Theft accomplished, the kid was about to run off when Julio did something unexpected.

'Hey, wait a minute,' he called after his mugger. 'If you're going to be robbing people for the rest of the night, you might as well take my coat to keep you warm.'

The boy turned back to Julio in disbelief. 'Why are you doing this?'

'If you're willing to risk your freedom for a few dollars,' Julio replied, 'then I guess you must really need the money. I mean, all I wanted to do was get dinner and if you really want to join me ... hey, you're more than welcome.'

The kid agreed, and moments later Julio and his assailant were seated at a booth in the diner. The waiters greeted them warmly. The manager stopped by for a chat. Even the dishwashers said hello.

'You know everybody here,' the kid said, surprised. 'Do you own this place?'

'No,' said Julio. 'I just eat here a lot.'

'But you're even nice to the dishwasher.'

'Well, haven't you been taught you should be nice to everybody?'

'Yeah,' the kid said, 'but I didn't think people actually behaved that way.'

When Julio and his mugger had finished eating, the bill arrived. But Julio no longer had his wallet. 'Look,' he told the kid. 'I guess you're going to have to pay for this bill because you have my money and I can't pay for this. So if you give me my wallet back, I'll gladly treat you.'

The kid gave him back his wallet. Julio paid the bill and then gave him $20. On one condition, he said: the teenager had to hand over his knife.

When a journalist later asked Julio why he'd treated his would-be robber to dinner, he didn't hesitate: 'I figure, you know, if you treat people right, you can only hope that they treat you right. It's as simple as it gets in this complicated world.'[1]

When I told a friend about Julio's act of kindness, he didn't miss a beat. 'Please excuse me while I barf.'

Okay, so this story is a little saccharine. It reminded me of the clichéd lessons I heard at church as a kid. Like the Sermon on the Mount, in Matthew 5:

> You have heard that it was said, 'An eye for an eye and a tooth for a tooth.' But I say to you, do not resist the one who is evil. But if anyone slaps you on the right cheek, turn to him the other also. And if anyone would sue you and take your tunic, let him have your cloak as well. And if anyone forces you to go one mile, go with him two miles.

Sure, you think. Swell plan, Jesus – if we were all saints. Problem is: we're all too human. And in the real world, turning the other cheek is about the most naive thing you can do. Right?

Only recently did I realise Jesus was advocating a quite rational principle. Modern psychologists call it *non-complementary behaviour*. Most of the time, as I mentioned earlier, we humans mirror each other. Someone gives you a compliment, you're quick to return the favour. Somebody says something unpleasant, and you feel the urge to make a snide comeback. In earlier chapters we saw how powerful these positive and negative feedback loops can become in schools and companies and democracies.

When you're treated with kindness, it's easy to do the right thing. Easy, but not enough. To quote Jesus again, 'If you love those who love you, what reward do you have? Do not even the tax collectors do the same? And if you greet only your brothers and sisters, what more are you doing than others?'[2]

The question is, can we take things a step further? What if we assume the best not only about our children, our co-workers, and our neighbours, but also about our enemies? That's considerably more difficult and can go against our gut instincts. Look at Mahatma Gandhi and Martin Luther King Jr, perhaps the two greatest heroes of the twentieth century. They were pros at non-complementary behaviour, but then again they were extraordinary individuals.

What about the rest of us? Are you and I capable of turning the other cheek? And can we make it work on a large scale – say, in prisons and police stations, after terrorist attacks or in times of war?

16

Drinking Tea with Terrorists

I

In a forest in Norway, about sixty miles south of Oslo, stands one of the strangest prisons in the world.

Here, you won't see cells or bars. You won't see guards armed with handguns or handcuffs. What you will see is a forest of birch and pine trees, and a rolling landscape crisscrossed by footpaths. Circling it all is a tall steel wall – one of the few reminders that people aren't here voluntarily.

The inmates of Halden prison each have a room of their own. With underfloor heating. A flatscreen TV. A private bathroom. There are kitchens where the inmates can cook, with porcelain plates and stainless steel knives. Halden also has a library, a climbing wall and a fully equipped music studio, where the inmates can record their own records. Albums are issued under their own label called – no joke – Criminal Records. To date, three of the prisoners have been contestants on Norway's *Idols*, and the first prison musical is in the works.[1]

Halden is a textbook example of what you might call a 'non-complementary prison'. Rather than mirroring the detainees' behaviour, staff turn the other cheek – even to hardcore felons. In fact, the guards don't carry weapons. 'We talk to the guys,' says one guard, 'that's our weapon.'[2]

If you're thinking this must be the softest correctional facility in Norway, you're wrong. Halden is a maximum-security prison.

And with some two hundred and fifty drug dealers, sexual offenders and murderers, it's also the second largest prison in the country.

If it's a softer prison you're after, that's just a couple miles up the road. A short drive away is Bastøy, a picturesque island that houses 115 felons who are sitting out the last years of their sentences. What happens here is analogous to the BBC's *Prison Experiment*, that yawn of a reality show that disintegrated into a pacifist commune (see Chapter 7).

When I first saw pictures of this island, I could scarcely believe my eyes. Inmates and guards flipping burgers together? Swimming? Lounging in the sun? To be honest, it was difficult to tell the prison staff apart from the inmates. Guards at Bastøy don't wear uniforms, and they all eat meals together, seated around the same table.

On the island, there are all kinds of things to do. There's a cinema, a tanning bed and two ski slopes. Several of the inmates got together and formed a group called the Bastøy Blues Band, which actually scored a spot opening for legendary Texas rockers ZZ Top. The island also has a church, a grocery store and a library.

Bastøy may sound more like a luxury resort, but it's not quite as laid-back as that. Inmates have to work hard to keep their community running: they have to plough and plant, harvest and cook, chop their own wood and do their own carpentry. Everything is recycled and they grow a quarter of their own food. Some inmates even commute off the island to jobs on the mainland, using a ferry service operated by the inmates themselves.

And, oh yes, for their work, the men also have access to knives, hammers and other potential weapons of murder. If they need to fell a tree, they can use a chainsaw. Even the convicted killer whose murder weapon was – you guessed it – a chainsaw.

Have the Norwegians lost it? How naive is it to sentence boatloads of murderers to a holiday resort? If you ask Bastøy's staff, nothing could be more normal. In Norway, where 40 per cent of prison guards are women, all guards have to complete a two-year training programme. They're taught that it's better to make friends with inmates than to patronise and humiliate them.

Norwegians call this 'dynamic security', to distinguish it from old-fashioned 'static security' – the kind with barred cells, barbed wire and surveillance cameras. In Norway, prison is not about preventing bad behaviour, but preventing bad intentions. Guards understand it to be their duty to prepare detainees, as best they can, for a normal life. According to this 'principle of normality', life inside the walls should resemble as closely as possible life on the outside.

And, incredibly, it works. Halden and Bastøy are tranquil communities. Whereas traditional penitentiaries are the quintessential *total institutions* – the kinds of places where bullying is rife (see Chapter 14) – in Norway's prisons the inmates get along fine. Anytime conflicts arise, both sides must sit down to talk it out, and they can't leave until they shake hands.

'It's really very simple,' explains Bastøy's warden, Tom Eberhardt. 'Treat people like dirt, and they'll be dirt. Treat them like human beings, and they'll act like human beings.'[3]

Even so, I still wasn't convinced. Rationally, I could understand why a non-complementary prison might work better. Intuitively, however, it seemed wrong-headed. How would it feel to victims, to know these murderers are being shipped off pleasantville?

But when I read Tom Eberhardt's explanation, it began to make sense. For starters, most inmates will be released sooner or later. In Norway, over 90 per cent are back on the streets in less than a year, so obviously they're going to be *somebody's* neighbour.[4] As Eberhardt explained to an American journalist, 'I tell people, we're releasing neighbours every year. Do you want to release them as ticking time bombs?'[5]

In the end, I reasoned, one thing matters more than anything else: the results. How do these kinds of prisons stack up? In the summer of 2018, a team of Norwegian and American economists got to work on this question. They looked at the *recidivism rate* – the chances someone will commit a repeat offence. According to the team's calculations, the recidivism rate among former inmates of penitentiaries like Halden and Bastøy is nearly 50 per cent lower than among offenders sentenced to community service or made to pay a fine.[6]

I was stunned. *Almost 50 per cent?* That's unheard of. It means that, for every conviction, on average eleven fewer crimes are committed in the future. What's more, the likelihood that an ex-convict will get a job is 40 per cent higher. Being locked up in a Norwegian prison really changes the course of people's lives.

It's no coincidence that Norway boasts the lowest recidivism rate in the world. By contrast, the American prison system has among the highest. In the US, 60 per cent of inmates are back in the slammer after two years, compared to 20 per cent in Norway.[7] In Bastøy it's even lower – a mere 16 per cent – making this one of the best correctional facilities in Europe, perhaps even the world.[8]

Okay, fine, but isn't the Norwegian method phenomenally expensive?

At the end of their 2018 article, the economists tallied the costs and benefits. A stay in a Norwegian prison, according to their calculations, costs on average $60,151 per conviction – almost twice as much as in the US. However, because these ex-convicts go on to commit fewer crimes, they also save Norwegian law enforcement $71,226 apiece. And because more of them find employment, they don't need government assistance and they pay taxes, saving the system on average another $67,086. Last but not least, the number of victims goes down, which is priceless.

Conclusion? Even using conservative estimates, the Norwegian prison system pays for itself more than two times over. Norway's approach isn't some naive, socialist aberration. It's a system that's better, more humane and less expensive.

2

On 23 July 1965, a commission of nineteen criminologists came together in Washington, D.C., convened by President Lyndon B. Johnson. Their assignment: over the next two years, develop a radical new vision for the American law enforcement system, spanning everything from policing to detention.

These were the turbulent 1960s. A new generation was pounding at the gates of power, crime rates were going up and the old criminal justice system was limping along. The criminologists knew it was time to think big. When they finally came out with their report, it contained more than two hundred recommendations. Emergency services needed an overhaul, police training needed to be stepped up and a national emergency hotline was needed – witness the birth of 911.

But the most radical recommendations concerned the future of US prisons. On this, the commission didn't mince words:

> Life in many institutions is at best barren and futile, at worst unspeakably brutal and degrading. [...] the conditions in which they [inmates] live are the poorest possible preparation for their successful reentry into society, and often merely reinforce in them a pattern of manipulation or destructiveness.[9]

It was time, said the commission, for total reform. No more bars, cells and endless hallways. 'Architecturally, the model institution would resemble as much as possible a normal residential setting,' advised the experts. 'Rooms, for example, would have doors rather than bars. Inmates would eat at small tables in an informal atmosphere. There would be classrooms, recreation facilities, dayrooms, and perhaps a shop and library.'[10]

It's a little-known fact that the United States nearly built a prison system similar to what Norway has today. Initial pilots with this 'new generation' of prisons were launched in the late sixties. In these facilities, detainees had rooms of their own, with doors opening onto a common area where they could talk, read and play games while an unarmed guard kept an eye on things. There was soft carpeting, upholstered furniture and real porcelain toilets.[11]

Behold, said the experts: the prison of the future.

In hindsight, it's shocking how fast the tide turned – and what caused it. It started with Philip Zimbardo, who in February 1973 published the first academic article on his Stanford Prison Experiment. Without having ever set foot in a real prison, the psychologist asserted that prisons are inherently brutal, no matter how you dress them up.

This verdict got a warm reception and gained popularity when the infamous Martinson Report appeared one year later. The man behind this report, Robert Martinson, was a sociologist at NYU with a reputation as a brilliant if slightly maniacal personality. He was also a man with a mission. In his younger years, Martinson had been a civil rights activist and landed in jail for thirty-nine days (including three in solitary confinement). This awful experience convinced him that all prisons are barbaric places.

In the late sixties, shortly after completing his degree, Martinson was invited to join a big project analysing a wide range of correctional strategies, from courses to therapy to supervision, aimed at helping criminals get on the right track. Working alongside two other sociologists, Martinson gathered data from more than two hundred studies done all over the world. Their final report, spanning 736 pages, was unimaginatively titled *The Effectiveness of Correctional Treatment: A Survey of Treatment Evaluation Studies.*

Since complex studies like this were rarely read by journalists, Martinson also published a short summary of their findings in a popular magazine. Title: 'What Works?' Conclusion: nothing works. 'With few and isolated exceptions,' Martinson wrote, 'the rehabilitative efforts that have been reported so far have had no appreciable effect on recidivism.'[12] The progressive social scientist hoped – much like Philip Zimbardo – that everyone would realise prisons were pointless places and should all be shut down.

But that's not what happened.

At first the media couldn't get enough of the charismatic sociologist. Newspapers and television programmes gave Martinson a platform to repeat his harsh verdict, while his co-authors stood by tearing their hair out. In reality, the results of 48 per cent of the studies analysed had been positive, showing rehabilitation can work.[13]

The skewed summary of the Martinson Report cleared the way for hardliners. Here was the proof, proclaimed conservative policymakers, that some people are simply born bad and stay bad. That the whole concept of rehabilitation defies human nature. Better to lock up these bad apples and throw away the key, they declared. This ushered in a new era of tough, tougher, toughest, and pulled the plug on America's experiment with a new generation of prisons.

Ironically, Martinson retracted his conclusion a couple years later ('contrary to my previous position, some treatment programs do have an appreciable effect on recidivism').[14] At a seminar in 1978, one astonished professor asked what he was supposed to tell his students. Martinson replied, 'Tell them I was full of crap.'[15]

By then, hardly anyone was listening. Martinson wrote one last article owning up to his mistakes, but only an obscure journal would run it. As another scientist observed, it was 'probably the most infrequently read article in the criminal justice debate on rehabilitation'.[16] Martinson's rectification passed unremarked by the newspapers, radio and TV. And it also didn't make the news when, a few weeks later, the fifty-two-year-old sociologist jumped from the fifteenth storey of his Manhattan apartment block.

3

By this point, someone else was making headlines: Professor James Q. Wilson.

His name may not ring any bells, but if we want to understand anything about how the US criminal justice system arrived at the state it's in today, there's no getting around this man. In the years after Robert Martinson took his own life, James Wilson would change the course of American history.

A political science professor at Harvard University, Wilson was the kind of guy with opinions about everything – from bioethics to the war on drugs and from the future of the constitutional state to scuba diving.[17] (He also loved being photographed with twenty-foot sharks).[18]

But the lion's share of his life's work centred on crime. If there was one thing Wilson hated, it was turning the other cheek. He had no use for the new generation of prisons that treated inmates with kindness. Exploring the 'origins' of criminal behaviour was a waste of time, he said, and all those liberals who yammered on about the effects of a troubled youth were missing the point. Some people are scum, pure and simple, and the best thing to do is lock them up. Either that or execute them.

'It is a measure of our confusion,' wrote the Harvard professor, 'that such a statement will strike many enlightened readers today as cruel, even barbaric.'[19]

To Wilson, however, it made perfect sense. His book *Thinking About Crime* (1975) became a big hit with top dogs in Washington, including President Gerald Ford, who in the year it was published called Wilson's ideas 'most interesting & helpful'.[20] Leading officials rallied around his philosophy. The best remedy against crime, Professor Wilson patiently instructed, was to put away the criminals. How hard could it be?

After reading a number of articles about James Q. Wilson's influence on the justice system, it hit me. I'd heard this name before.

Turns out that in 1982 Wilson came up with another revolutionary idea, which would enter the history books as the 'broken windows' theory. The first time I encountered this theory was in the same book where I'd also first read about Kitty Genovese's murder (and the thirty-eight bystanders): journalist Malcolm Gladwell's *The Tipping Point*.

I remember being riveted by his chapter on Wilson. 'If a window in a building is broken and is left unrepaired,' Wilson wrote in a piece for *The Atlantic* in 1982, 'all the rest of the windows will soon be broken.'[21] Sooner or later, if nobody intervenes, the vandals will be followed by squatters. Next, drug addicts might move in, and then it's only a matter of time before someone gets murdered.

'This is an epidemic theory of crime,' Gladwell observed.[22] Litter on the sidewalks, vagrants on the street, graffiti on the walls: they're all precursors to murder and mayhem. Even a single broken window sends the message that order is not being enforced, signalling to criminals that they can go even further. So, if you want to fight serious crime, you have to start by repairing broken windows.

At first, I didn't get it. Why worry about minor offences when people are getting murdered every day? It sounded – Gladwell conceded – 'as pointless as scrubbing the decks of the *Titanic* as it headed toward the icebergs'.[23]

But then I read about the first experiments.

In the mid-1980s, New York City's subways were covered in graffiti. The Transit Authority decided something needed to be done, so they hired Wilson's co-author George Kelling as a consultant. He recommended a large-scale clean-up. Even trains with just a little graffiti were swiftly sent off to be scrubbed clean. According to the subway director, 'We were religious about it.'[24]

Then came phase two. Wilson and Kelling's broken windows theory applied not only to disorder, but to the people who cause it. A city where beggars, hoodlums and panhandlers were allowed to roam at will was setting itself up for much worse. After all, as Wilson noted in 2011, 'public order is a fragile thing'.[25] Unlike many other scientists, he put little stock in investigating the structural causes of crime, such as poverty or discrimination. Instead he stressed that there's ultimately only one cause that matters. Human nature.

Most people, Wilson believed, do a simple cost-benefit calculation: does crime pay or not? If police are lax or jails too comfortable, it's a sure bet more people will choose a life of crime.[26] If crime rates rise, the solution is equally straightforward.

You fix it with stronger extrinsic incentives like higher fines, longer jail time and harsher enforcement. As soon as the 'costs' of crime go up, demand will drop.

One man couldn't wait to put Wilson's theory into practice: William Bratton. And Bratton is the final linchpin in our story. In 1990, he was appointed the new chief of the New York City Transit Police. Bratton was a fervent believer in James Wilson's doctrine. An energetic man, he was famous for constantly handing out copies of the original broken windows article published in *The Atlantic*.

But Bratton wanted to do more than repair windows. He wanted to restore order in New York City, with an iron fist. As his first target, he picked fare evaders. Subway users who couldn't present a $1.25 ticket from now on were arrested by transit cops, handcuffed and ceremoniously lined up on the subway platform where everyone could get a good look. The number of arrests made quintupled.[27]

This merely whetted Bratton's appetite. In 1994, he was promoted to city police commissioner, and soon all New Yorkers were getting a taste of Bratton's philosophy. Initially, his officers were obstructed by rules and protocols, but Bratton swept them away. Now, anyone could be arrested for even the slightest infraction – public drinking, getting caught with a joint, joking around with a cop. In Bratton's own words, 'If you peed in the street, you were going to jail.'[28]

Miraculously, this new strategy appeared to work. Crime rates *plummeted*. Murder rate? Down 63 per cent between 1990 and 2000. Muggings? Down 64 per cent. Car theft? Down 71 per cent.[29] The broken windows theory once ridiculed by journalists turned out to be a stroke of genius.

Wilson and Kelling became the country's most esteemed criminologists. Commissioner Bratton made it onto the cover of *Time* magazine and went on to be appointed chief of the Los Angeles Police Department in 2002 and reappointed by the NYPD in 2014. He came to be revered by generations of police officers, who styled themselves 'Brattonistas'.[30] Wilson even credited Bratton with the 'biggest change in policing in the country'.[31]

4

Almost forty years have passed since the broken windows article first ran in *The Atlantic*. During that time, Wilson and Kelling's philosophy has percolated into the farthest reaches of the United States, and well beyond, from Europe to Australia. In *The Tipping Point*, Malcolm Gladwell calls the theory a great success, and in my first book I was enthusiastic about it, too.[32]

What I failed to realise was that by then few criminologists believed in it any longer. Actually, alarm bells should have started ringing as soon as I read in *The Atlantic* that Wilson and Kelling's theory was based on one dubious experiment.

In this experiment, a researcher left a car parked in a respectable neighbourhood for a week. He waited. Nothing happened. Then he returned with a hammer. No sooner had the researcher himself smashed one car window than the floodgates opened. Within a matter of hours, ordinary passers-by had demolished the car.

The researcher's name? Philip Zimbardo!

Zimbardo's car experiment, never published in any scientific journal, was the inspiration for the broken windows theory. And just like his Stanford Prison Experiment, this theory has since been thoroughly debunked. We know, for instance, that the 'innovative' policing of William Bratton and his Brattonistas was not responsible for the drop in New York City's crime rates at all. The decline set in earlier, and in other cities, too. Cities like San Diego, where the police left minor troublemakers alone.

In 2015, a meta-analysis of thirty studies on broken windows theory revealed that there's no evidence Bratton's aggressive policing strategies did anything to reduce crime.[33] Zip, zero, zilch. Neighbourhoods aren't made safer by issuing parking tickets, just as you couldn't have saved the *Titanic* by scrubbing the deck.

My initial reaction was: okay, so arresting bums and drunks doesn't reduce serious crime. But it's still good to enforce public order, right?

This throws up a fundamental question. Whose 'order' are we talking about? Because as arrests in the Big Apple skyrocketed, so did the reports of *police* misconduct. By 2014, thousands of demonstrators were taking to the streets of New York and other US cities, including Boston, Chicago and Washington. Their slogan: 'Broken windows, broken lives.'

This was no exaggeration. In the words of two criminologists, aggressive policing was leading to citations for:

> ... women eating doughnuts in a Brooklyn park; chess players in an Inwood park; subway riders for placing their feet on seats at 4am and an elderly Queens couple cited for no seatbelts on a freezing cold night while driving to purchase needed prescription drugs. Allegedly, the man was instructed to walk home to secure identification – a few blocks from the pharmacy. When he returned to the pharmacy, the officers already wrote the ticket using a prescription bottle as identification. The elderly man's subsequent heart attack led to his death.[34]

What had sounded so good in theory boiled down to more and more frivolous arrests. Commissioner Bratton became obsessed with statistics, and so did his officers. Those who could present the best figures were promoted, while those who lagged behind were called to task. The result was a quota system in which officers felt pressured to issue as many fines and rack up as many citations as possible. They even began fabricating violations. People talking in the street? Arrest 'em for blocking a public road. Kids dancing in the subway? Book 'em for disturbing the peace.

Serious crimes were an entirely different story, investigative journalists later discovered. Officers were pressured to tone down their reports or skip them altogether, to avoid making departmental figures look bad. There are even cases of rape victims who were subjected to endless questioning in an attempt to trip them up on tiny inconsistencies. Then the incident wouldn't be included in the data.[35]

On paper, it all looked fantastic. Crime had taken a nose-dive, the number of arrests were sky high and Commissioner Bratton was the hero of New York. In reality, criminals walked free while thousands of innocent people became suspects. To this day, police departments across the country still swear by Bratton's philosophy – which is why scientists continue to consider US police statistics unreliable.[36]

There's more. The broken windows strategy has also proven synonymous with racism. Data show that a mere 10 per cent of people picked up for misdemeanours are white.[37] Meanwhile, there are black teenagers who get stopped and frisked on a monthly basis – as they have been for years – despite never having committed an offence.[38] Broken windows has poisoned relations between law enforcement and minorities, saddled untold poor with fines they can't pay and also had fatal consequences, as in the case of Eric Garner, who died in 2014 while being arrested for allegedly selling loose cigarettes. 'Every time you see me, you want to mess with me,' Garner protested. 'I'm tired of it ...

Please just leave me alone. I told you the last time, please just leave me alone.'

Instead, the officer wrestled him to the ground and put him in a chokehold. Garner's last words were 'I can't breathe.'

Only now, years after reading Malcolm Gladwell's book, have I come to realise the broken windows theory is underpinned by a totally unrealistic view of human nature. It's yet another variant on veneer theory. It made police in New York treat ordinary people like potential criminals: the smallest misstep could supposedly be the first on a path to far worse. After all, our layer of civilisation is tenuously thin.

Officers, meanwhile, were being managed as though they possessed no judgement of their own. No intrinsic motivation. They were drilled by their superiors to make their departments look as good as possible on paper.

Does this mean we should forget about fixing actual broken windows? Of course not. It's an excellent idea to repair windows, spruce up houses and listen to local people's concerns. Just as an orderly prison radiates trust, a tidy neighbourhood feels much safer.[39] And after you fix the windows, you can throw them wide open.

But Wilson and Kelling's argument wasn't primarily about broken windows or poorly lit streets. The 'broken window' was a misleading metaphor. In practice, it was ordinary people who were being registered, restrained and regulated.

Professor Wilson was steadfast to the end, maintaining right up to his death in 2012 that the Brattonista approach was a huge success. Meanwhile, his co-author was plagued by mounting doubts. George Kelling felt the broken windows theory had been too often misapplied. His own concern had always been about the broken windows themselves, not the arrest and incarceration of as many black and brown people as possible.

'There's been a lot of things done in the name of Broken Windows that I regret,' Kelling admitted in 2016. When he began hearing police chiefs all over the country invoke his theory, two words flashed across his mind: 'Oh s—t.'[40]

What would happen if we turned the broken windows theory around? If we can redesign prisons, could we do the same with police departments?

I think we can. In Norway – where else – there's already a long tradition of community policing, a strategy that assumes most folks are decent, law-abiding citizens. Officers work to win community trust, informed by the idea that if people know you, they'll be more likely to help out. Neighbours will give more tips, and parents will be quicker to call if their child seems to be heading down the wrong path.

Back in the 1970s, Elinor Ostrom – the economist who researched the commons (see Chapter 15) – conducted the largest study ever into police departments in the United States. She and her team discovered that smaller forces invariably outperform bigger ones. They're faster on the scene, solve more crimes, have better ties with the neighbourhood, and all at a lower cost. Better, more humane, less expensive.[41]

In Europe, the philosophy of community policing has been around a while. Officers are used to coordinating with social services and even consider what they do a kind of social work.[42] They're also well trained. In the US, the average police training programme lasts just nineteen weeks, which is unthinkable in most of Europe. In countries like Norway and Germany, law enforcement training takes more than two years.[43]

But some American cities are changing their approach. The people of Newark, New Jersey, elected a new, black mayor in 2014, and he had a clear vision of what modern policing in the city should look like. It requires officers, he said, 'who know people's grandmothers, who know the institutions of the community, who look at people as human beings [...] that's the beginning of it. If you don't look at the people you're policing as human, then you begin to treat them inhumanely.'[44]

Could we take the principle of turning the other cheek even further? Absurd as the question may sound, I couldn't help wondering: Could a non-complementary strategy also work in the war on terror?

In my search for an answer I discovered that this strategy has been tried already – in fact, in my own country. Among experts, it's even called the Dutch Approach. It began back in the seventies, when the Netherlands was confronted with a violent wave of leftist terrorism. The government didn't enact new security laws, however, and the media did as law enforcement asked and limited coverage. While West Germany, Italy and the United States brought out the big guns – helicopters, road blocks, troops – the Netherlands refused to give terrorists the platform they wanted.

In fact, the police refused to even use the word 'terrorism'. They preferred 'violent political activism' or plain old 'criminals'. Meanwhile, Dutch intelligence was busy behind the scenes, infiltrating extremist groups. They set their sights specifically on terrorists – sorry, criminals – without turning whole segments of the population into suspects.[45]

This led to some comical situations, like a tiny Red Youth cell in which three of the four members were undercover agents. Carrying out attacks turned out to be pretty difficult when there was always someone off taking a bathroom break or holding the map upside down.

'A behind-the-scenes, timely, cautious counterterrorism policy,' concludes a Dutch historian, 'brought the spiralling violence to a standstill.'[46] When some of the Red Youth visited a terrorist training camp in Yemen, the Dutch terrorists were shocked by the intensity of the German and Palestinian fighters. It was downright scary. As one Dutch member later put it, 'They took all the fun out of it.'[47]

A more recent example of a turn-the-other-cheek approach comes from the Danish city of Aarhus. In late 2013, the police decided not to arrest or jail young Muslims who wanted to go and fight in Syria but to offer them a cup of tea instead. And a mentor. Family and friends were mobilised to make sure these teenagers knew there were people who loved them. At the same time, police strengthened ties with the local mosque.

More than a few critics called the Aarhus approach weak or naive. But in truth, the police chose a bold and difficult strategy. 'What's easy,' scoffed the police superintendent, 'is to pass tough new laws. Harder is to go through a real process with individuals: a panel of experts, counselling, healthcare, assistance getting back into education, with employment, maybe accommodation [...] We don't do this out of political conviction; we do it because we think it works.'[48]

And work it did. While in other European cities the exodus continued unabated, the number of jihadists travelling to Syria from Aarhus declined from thirty in 2013 to one in 2014 and two in 2015. 'Aarhus is the first, to my knowledge, to grapple with [extremism] based on sound social psychology evidence and principles,' notes a University of Maryland psychologist.[49]

And then there's Norway. People there managed to keep a cool head even after the most horrific attack in the country's history. After the 2011 bloodbath perpetrated by right-wing extremist Anders Breivik, the country's prime minister declared, 'Our response is more democracy, more openness, and more humanity.'[50]

This kind of response often gets you accused of looking the other way, of choosing the easy path. But more democracy, more openness and more humanity are precisely what's *not* easy. On the contrary, tough talk, retaliation, shutting down borders, dropping bombs, dividing up the world into the good guys and the bad – that's easy. *That's* looking the other way.

5

There are some moments when it becomes impossible to look away. When the truth refuses to be ignored. In October 2015, a delegation of North Dakota's top prison officials experienced just such a moment.

It happened during a work visit to Norway. For those who don't know, North Dakota is a sparsely populated, conservative state. The incarceration rate is eight times as high as in Norway.[51] And the prisons? They're old-fashioned holding pens; all long corridors, bars and stern guards. The American officials didn't expect to learn much from their trip. 'I was arrogant,' one said later. 'What was I really going to see other than what I call the IKEA prison?'[52]

But then they saw the prisons. Halden. Bastøy. The tranquillity. The trust. The way inmates and guards interact.

Seated at the bar of the Radisson Hotel in Oslo one evening was the director of the North Dakota Department of Corrections. Leann Bertsch – known among her colleagues as tough and unbending – began to cry. 'How did we think it was okay to put human beings in cagelike settings?'[53]

Between 1972 and 2007, the number of people incarcerated in the United States, corrected for population growth, grew by more than 500 per cent.[54] And those inmates are locked up for an average of sixty-three months – seven times longer than in

Norway. Today, almost a quarter of the world's prison population is behind American bars.

This mass incarceration is the result of intentional policy.
The more people you lock up, Professor James Wilson and
his followers believed, the lower the crime rate. But the truth
is that many American prisons have devolved into training
grounds for criminals – costly boarding schools that produce
more accomplished crooks.[55] A few years ago, it came out that a
mega-facility in Miami was cramming as many as twenty-four
inmates into a single cell, from which they were let out for one
hour twice a week. The result was a 'brutal gladiatorial code of
fighting' among inmates.[56]

Individuals released from these kinds of institutions are a
genuine danger to society. 'The vast majority of us become
exactly who we are told we are,' says one former California
prison inmate: 'violent, irrational, and incapable of conducting
ourselves like conscious adults.'[57]

When Leanne Bertsch returned from Norway, she realised
things had to change in North Dakota prisons. She and her team
formulated a new mission: 'To implement our humanity.'[58]

Step one? Shelve the broken windows strategy. Where
before there had been regulations covering over three hundred
violations – not tucking in your shirt, for instance, which could
land you in solitary confinement – now all those nitpicky rules
were scrapped.

Next, a new protocol was drawn up for guards. Among other
things, they had to have at least two conversations a day with
inmates. This was a major transition and met with considerable
resistance. 'I was scared to death,' one of the guards recalled.
'I was scared for staff. I was scared for the facility. I was scared
when we talked about specific guys leaving, and I was wrong.'[59]

As the months passed, the guards began to take more pleasure in their work. They started a choir and painting classes. Staff and inmates began playing basketball together. And there was a notable reduction in the number of incidents. Before, there had been incidents 'at least three or four times a week', according to one guard. 'Someone trying to commit suicide, or someone trying to flood their cell, or being completely disorderly. We haven't hardly had any of that this year.'[60]

Top officials from six other US states and counting have since taken trips to Norway. Director Bertsch in North Dakota continues to stress that reform is a matter of common sense. Locking away whole swaths of the population is just a bad idea. And the Norwegian model is demonstrably better. Less expensive. More realistic.

'I'm not a liberal,' swears Bertsch. 'I'm just practical.'[61]

17

The Best Remedy for Hate, Injustice and Prejudice

1

I couldn't stop thinking about the idea behind the Norwegian prisons. If we can turn the other cheek with criminals and would-be terrorists, then maybe we can apply the same strategy on a larger scale. Maybe we can bring together sworn enemies and even stamp out racism and hatred.

I was reminded of a story I'd come across in a footnote somewhere, but hadn't pursued. A tale of two brothers who for decades stood on opposing sides, yet in the end managed to prevent a full-blown civil war. Sounds like a good story, doesn't it? In a pile of old notes, I found the brothers' names, and after that, I wanted to know everything about them.

2

The story of the brothers is inextricably bound up with one of the most renowned figures of the twentieth century. On 11 February 1990, millions of people sat glued to their televisions to see him. Nelson Mandela, imprisoned for twenty-seven years, became a free man on that day. Finally, there was hope for peace and reconciliation between black and white South Africans. 'Take your guns, your knives and your pangas,' shouted Mandela shortly after his release, 'and throw them into the sea!'[1]

Four years later, on 26 April 1994, the first elections were held for *all* South Africans. Again the images were enthralling: endless lines at the polling stations, twenty-three million voters in all. Black men and women old enough to remember the start of apartheid casting ballots for the first time in their lives. Helicopters that once brought death and destruction, now dropping pencils and paper ballots.

A racist regime had fallen and a democracy was born. Two weeks later, on 10 May, Mandela was sworn in as the country's first black president. During his inauguration, fighter jets flashed across the sky tracing vapour trails in the colours of the Rainbow Nation. Combining green, red, blue, black, white and gold, the new South African flag was the most colourful on earth.

Less well known: how close it came to not happening at all.

The South Africa we know today nearly didn't make it. In the four years between Mandela's release and his election as president, the country came to the brink of civil war. And wholly forgotten is the crucial role two brothers – identical twins – had in preventing it.

Constand and Abraham Viljoen were born on 28 October 1933. As boys they were inseparable.[2] The brothers attended the same schools and were in the same classes. They listened to the same teachers and the same propaganda about the superiority of the white race.

More importantly, they were moulded by the same history. Constand and Abraham were Afrikaners. They were the descendants of French Huguenots who came ashore in 1671 and intermingled with the Dutch settlers. In 1899, this Afrikaner population would rise up against British rule in South Africa, only to be ruthlessly crushed.

The boys' father had experienced the British concentration camps as a child. He'd looked on, helpless, as his brother and two sisters died in their mother's arms. Constand and Abraham's family thus belonged to an oppressed people, but sometimes

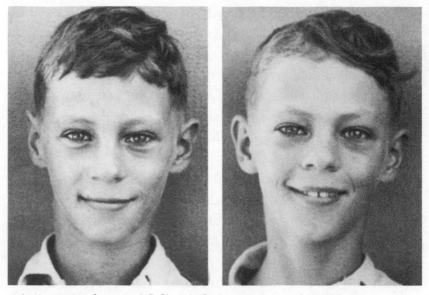

School portraits of Constand (left) and Abraham. Source: Andries Viljoen.

the oppressed become the oppressors, and it was this truth that would drive the twins apart.

In 1951, soon after the boys' eighteenth birthday, their mother announced there wasn't enough money to send them both to college in Pretoria. You go ahead, Constand said to Abraham, or 'Braam' as he was known. After all, Braam was the smart one.

While his brother enrolled in theology, Constand enlisted in the military. Army life suited him, and it became like a second family. While Braam pored over his books, Constand jumped out of helicopters. While Braam studied in Holland and America, Constand fought in Zambia and Angola. And while Braam befriended students from all over the world, Constand developed a deep bond with his military comrades.

Year by year, the brothers drifted further apart. 'I was exposed to the question of just treatment,' Braam later recalled, 'and to the belief that people were equals.'³ Braam began to realise that the apartheid he'd grown up with was a criminal system and contradicted everything the Bible taught.

When he returned after years of studying abroad, many South Africans considered Braam a deserter. A heretic. A traitor. 'They said I had been influenced,' he said later. 'That I never should have been allowed to go overseas.'⁴ But Braam wasn't dissuaded and continued to call for the equal treatment of his black coun- trymen. In the eighties he ran for office, representing a party that sought to end apartheid. It became increasingly clear to him that the apartheid government was a downright murderous regime.

Constand, meanwhile, grew to be one of South Africa's most beloved soldiers. His uniform was soon spangled with medals. At the pinnacle of his career, he became chief of the South African Defence Force, encompassing the army, navy and air force. And until 1985 he remained apartheid's great champion.

In time, the Viljoen brothers stopped speaking altogether. Hardly anybody remembered that General Viljoen – the pat- riot, war hero and darling of scores of Afrikaners – even had a twin brother.

Yet their bond would determine the future of South Africa.

3

How do you reconcile sworn enemies?

With that question in mind, an American psychologist set out for South Africa in the spring of 1956. Apartheid had already been imposed. Mixed marriages were prohibited and later that year the administration would adopt a law reserving better jobs for whites.

The psychologist's name was Gordon Allport, and all his life he'd pondered two basic questions: 1) Where does prejudice come from, and 2) How can you prevent it? After years of research, he'd found a miracle cure. Or at least he thought he had.

What was it?

Contact. Nothing more, nothing less. The American scholar suspected that prejudice, hatred and racism stem from a lack of contact. We generalise wildly about strangers because we don't know them. So the remedy seemed obvious: more contact.

Most scientists were not impressed and called Allport's theory simplistic and naive. With the Second World War still fresh in people's minds, the general consensus was that more contact led to *more* friction. In those very same years, psychologists in South Africa were still investigating the 'science' of differences in racial biology that would justify 'separate development' (read: apartheid).[5]

For many white South Africans, Allport's theory was positively shocking. Here was a scientist arguing that apartheid wasn't the solution to their problems, but the cause. If blacks and whites could only meet – at school, at work, in church, or anywhere at all – they could get to know one another better. After all, we can only love what we know.[6]

This, in a nutshell, is the contact hypothesis. It sounds too simple to be believed, but Allport had some evidence to back it up. He pointed to the race riots that broke out in Detroit in 1943, for instance, where sociologists had noticed something strange: 'People who had become neighbors did not riot against each other. The students of Wayne University – white and black – went to their classes in peace throughout Bloody Monday. And there were no disorders between white and black workers in the war plants ...'[7]

On the contrary, people who were neighbours had shielded one another. Some white families sheltered their black neighbours when rioters came around. And vice versa.

Even more remarkable were the data gathered by the US military during the Second World War. Officially, black and white soldiers were not supposed to fight side by side, but in the heat of battle it sometimes happened. The army's research office discovered that in companies with both black and white platoons, the number of white servicemen who disliked blacks was far lower. To be precise, *nine times* lower.[8]

Gordon Allport wrote page after page about the positive effects of contact. It applied to soldiers and police officers, to neighbours and students. If black children and white children attended the same schools, for example, they were seen to lose their prejudices. This meant that what Braam Viljoen experienced during his studies abroad wasn't exceptional. It was the rule.

Perhaps the most powerful proof for Allport's contact hypothesis came from the sea. When African Americans were first admitted to the largest seamen's union in 1938, there was initially widespread resistance. But once black and white seamen actually began working together, the protests ceased.[9]

The best remedy for racism? Sailing together.

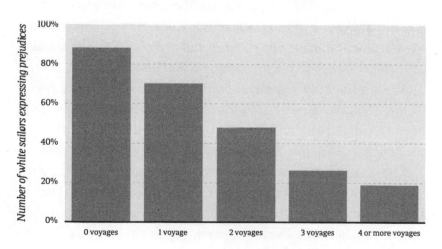

Number of voyages with a mixed crew

Gordon Allport was a cautious man; he knew his case was still far from watertight. It's possible that the sailors who signed on for mixed crews might be less racist to begin with.

As he travelled through South Africa in 1956 – two years after publishing his magnum opus on contact theory – Allport's initial doubts resurfaced.[10] In this country where blacks and whites had been living side by side for centuries, racism was not diminishing. If anything, it seemed to be increasing. Of the many white Afrikaners Allport met, none seemed to have mental disorders, yet all continued to exclude and discriminate. So did his theory really hold up?

Looking back in the sixties on his visit to South Africa, Allport felt forced to concede that he'd been blind to 'the forces of history'.[11]

4

It's 7 May 1993. Crowded inside the rugby stadium in Potchefstroom, about seventy-five miles south of Johannesburg, are 15,000 white Afrikaners. Above them wave hundreds of red and black flags bearing symbols that resemble nothing so much as swastikas. Sporting long beards and brown shirts, the farmers are armed to the teeth with shotguns and pistols.[12]

Among the rally's speakers is Eugène Terre'Blanche, leader of the Afrikaner Resistance Movement. Terre'Blanche has long been fascinated by the oratorical techniques of Adolf Hitler and his henchmen are like the Ku Klux Klan, only more violent.

That day the stadium seethes with anger and fear. Fear of what will happen if Mandela wins the first all-race elections. Fear of losing their national flag and anthem. Fear of the obliteration of an entire culture. These 15,000 angry demonstrators are also known as *Bittereinders*, after the Afrikaners who a hundred years earlier fought the British to the bitter end. They see themselves as freedom fighters, prepared to use whatever means necessary.

Only, they're missing something; or, rather, someone. What they need now is a leader. Someone who commands respect. Someone with an exemplary track record. Someone who can be for the Afrikaners what Mandela is for the 'black danger' – the *swart gevaar* – and who will lead them in this final, momentous battle for freedom.

Someone, in short, like Constand Viljoen.

Constand is there that day in Potchefstroom. Having retired years before, he now leads a quiet life as a farmer. But when the mob begins chanting his name, he doesn't hesitate. The former general takes the stage.

'The Afrikaner people must prepare to defend themselves,' Constand roars into the microphone. 'A bloody conflict which requires sacrifices is inevitable, but we will gladly sacrifice because our cause is just!'

The crowd goes wild.

'You lead,' cry the Afrikaners, 'we will follow!'[13]

This is how Constand becomes the leader of a new coalition calling itself the *Afrikaner Volksfront*. And this isn't just any political party or federation. It's an army. Constand is mobilising for war. He wants to prevent multiracial elections at all costs.

'We had to build a massive military capability,' Constand later recalls.[14] In two short months, the AVF recruits 150,000 Afrikaners, including 100,000 experienced servicemen. The mere mention of the name 'Constand Viljoen' is enough to convince most.

At the same time, they need to devise a plan of attack, which leads to a succession of crackbrained proposals. Maybe they should ambush the leadership of the ANC, Mandela's political party, one person suggests. No, says another, they should lynch 15,000 black people in Western Transvaal and dump them in a mass grave. With each passing day, the mood grows more rabid.

Seventy-five miles away in Johannesburg, Constand's brother Abraham feels a deep sense of foreboding. 'Sometimes I think that the classic elements of tragedy are constellating here,' he writes in a memo to Mandela and the ANC.[15] But Braam also realises he needs to act. He knows he's the only person in the whole of South Africa who may be able to change his brother's mind. After hardly talking in forty years, now they must.

'If he could win Constand over,' one historian would later write, 'a peaceful transition from apartheid to democracy would be possible. If not, war was inevitable.'[16]

It's early July 1993, with ten months until the elections, when Braam arrives at the AVF offices in downtown Pretoria.

As soon as the two brothers are seated, Braam cuts to the chase.

'What are your options?'

'As things stand now,' Constand replies, 'we have only one option, and that is to fight.'[17]

Then Braam makes a proposal, a plan he and Nelson Mandela have hammered out together in the utmost secrecy. What would Constand say, Braam asks, to sitting down with the ANC leadership for direct talks about the position of his people? By this point, Constand has already rejected nine such overtures. But this time his response is different.

This time it's his brother asking.

And so it transpires that a pair of identical twins arrive together on the doorstep of a villa in Johannesburg on 12 August 1993. They expect to be greeted by household staff, but standing before them with a big grin is the man himself. Nelson Mandela.

It's a historic moment: the hero of the new South Africa standing eye to eye with the hero of the old. The peacemaker opposite the man mobilising for war. 'He asked me if I took tea,' Constand remembered years after the event. 'I said yes and he poured me a cup. He asked me if I took milk. I said yes and he poured me milk. Then he asked me if I took sugar with my tea. I said I did and he poured the sugar. All I had to do was stir it!'[18]

As they talk, it's obvious that Mandela has made an effort to understand the history and culture of the Afrikaners. Constand is impressed when Mandela draws parallels between the Viljoen family's struggle for freedom from the British a hundred years ago and his own fight against apartheid. Most important, historians later note, is that Mandela talks to the military man in his own language. 'General,' he says in Afrikaans, 'there can be no winners if we go to war.'

Constand nods, 'There can be no winners.'[19]

That first meeting opens four months of secret talks between Viljoen and Mandela. Even President Frederik Willem de Klerk is kept in the dark, and few history books mention it today. Yet this was a pivotal moment in the history of South Africa. In the end, the former general was convinced to lay down his weapons and join the elections with his party.

Each time Constand shook Mandela's hand, his admiration grew for the man he once considered a terrorist. And the feeling was mutual. Mandela developed a growing respect for the general, and unlike career politician De Klerk, came to trust him.

'He took my brother's arm,' Braam would later say, 'and he did not let it go.'[20]

5

By then Gordon Allport, the psychologist who had come up with the contact hypothesis, was long dead. But the student he'd toured South Africa with in 1956 was still very much alive.

Unlike the retiring Allport, Thomas Pettigrew was a rebel. An activist. He'd played a visible role in the American civil rights movement and the FBI had a thick file on his activities. While staying in South Africa, Pettigrew attended a string of illegal ANC meetings, and the secret service had taken note. When he presented his passport at customs six months later, it got a stamp in big letters: 'BANNED FROM SOUTH AFRICA.'[21]

Little did Pettigrew know that he would one day return to the land of Mandela. Half a century later, in 2006, he was invited to an international psychology conference in South Africa.

'Everywhere we looked,' said Pettigrew of his trip, 'we saw progress even though much remains to be accomplished.'[22] Durban's beautiful beaches were now open to all. On the spot once occupied by an infamous prison, the Constitutional Court now stood with a sign welcoming visitors in South Africa's eleven official languages.

As one of the leading scientists in his field and the convention's guest of honour, Pettigrew presented a massive study that provided overwhelming support for his former mentor's theory. Pettigrew and his team rounded up and analysed 515 studies from thirty-eight countries.[23] Their conclusion? Contact *works*. Not only that, few findings in the social sciences have this much evidence to back them up.

Contact engenders more trust, more solidarity and more mutual kindness. It helps you see the world through other people's eyes. Moreover, it changes you as a person, because individuals with a diverse group of friends are more tolerant towards strangers. And contact is contagious: when you see a neighbour getting along with others, it makes you rethink your own biases.

But what also came out of these studies was the finding that a single negative experience (a clash or an angry look) makes a deeper impression on us than a joke or a helping hand. That's just how our brains work. Initially, this left Pettigrew and his colleagues with a puzzle. Because if we have a better memory for bad interactions, how come contact nonetheless brings us closer together? The answer, in the end, was simple. For every unpleasant incident we encounter, there are any number of pleasant interactions.[24]

The bad may seem stronger, but it's outnumbered by the good.

If there's one person who understood the power of contact it was Nelson Mandela. Years earlier, he had chosen a very different path – the path of violence. In 1960, Mandela had been one of the founding members of the armed wing of the ANC.

But twenty-seven years behind bars can utterly change a person. As the years passed, Mandela began to realise what scientists would later show: nonviolent resistance is a lot more effective than violence. Take the recent work of Erica Chenoweth, an American sociologist who started out believing the 'Mandela Method' was naive. In the real world, she thought, power is exercised through the barrel of a gun. To prove it, she created a huge database of resistance movements going back to 1900.

'Then I ran the numbers,' she wrote in 2014. 'I was shocked.'[25] More than 50 per cent of the nonviolent campaigns were successful, as opposed to 26 per cent of the militant ones. The primary reason, Chenoweth established, is that more people join nonviolent campaigns. On average over *eleven times* more.[26] And not just guys with too much testosterone, but also women and children, the elderly and people with disabilities. Regimes just aren't equipped to withstand such multitudes. That's how good overpowers evil – by outnumbering it.

In nonviolent campaigns, one ingredient is essential: self-control. While in prison, Mandela became a master at keeping a cool head. He decided to study his enemy, reading scores of books about the culture and history of the Afrikaners. He watched rugby. He learned their language. 'If you talk to a man in a language he understands,' Mandela explained, 'that goes to his head. If you talk to him in his language, that goes to his heart.'[27]

Mandela tried to make his fellow inmates see that their guards were people, too, only that they'd been poisoned by the system. Years later, that's how Mandela would look at Constand Viljoen: as an honest, loyal and brave man who'd spent his life fighting for a regime he believed in.

After his release, Mandela was able to rally 90 per cent of black South Africans to the cause. He then turned his efforts to winning the hearts of white Afrikaners. Such was his success that when Mandela entered Johannesburg's stadium dressed in the white rugby team's shirt on 24 June 1995, he was greeted with cheers of 'Nelson, Nelson!' by thousands of men and women who'd once thought him a terrorist.

It's tempting to attribute the triumph of Mandela's approach to a gift for publicity, but that's not what did it. He didn't orate with the passion of a Martin Luther King or debate with the fire of a Winston Churchill. At his first press conference he was bewildered by the furry objects clustered in front of him, until someone whispered in his ear that they were microphones.[28]

Mandela's superpower lay elsewhere. What made him one of the greatest leaders in world history, observes journalist John Carlin, is that 'he chose to see good in people who ninety-nine people out of a hundred would have judged to have been beyond redemption'.[29]

Walter Sisulu, one of Mandela's closest friends, was once asked to name some of Mandela's flaws. 'When he trusts a person,' Sisulu began, 'he goes all out ...'

Then he hesitated.

'But perhaps it is not a failing ...'[30]

6

Looking back on the most hopeful shifts in recent decades, we see that trust and contact were instrumental every time. Take the emancipation of gays and lesbians starting from the 1960s. As more and more brave souls came out of the closet, friends and co-workers and mothers and fathers learned that not everybody has the same sexual preference. And that that's okay.

But the opposite also holds true. After Donald Trump was elected president in 2016, it became clear that all too often we still live in our own bubbles. Two sociologists even showed that 'the racial and ethnic isolation of Whites at the zip-code level is one of the strongest predictors of Trump support'.[31] And also that the *further* the distance to the border separating the US from Mexico, the higher the support for the man who campaigned on building a giant wall between the two.[32] The problem, in other words, was not too much contact between Trump voters and Muslims and refugees, but too little.

The same pattern played out in the referendum held in Britain in 2016 on whether to leave the EU. In communities that were less culturally diverse, proportionately more residents voted in favour of Brexit.[33] And in my own country of Holland, the highest concentrations of populist party voters are found in areas with the highest concentrations of white residents. A Dutch team of sociologists found that when whites had more contact with Muslims (primarily at work), they were also less Islamophobic.[34]

Not only that, diversity can also make us friendlier. In 2018, an international team of researchers at the University of Singapore established on the basis of five new studies that people who live in more diverse communities more often identify with all of humanity. As a result, they also exhibit more kind, helpful behaviour towards strangers. This was demonstrated after the

Boston Marathon bombing in 2013, when residents of more diverse neighbourhoods provided more help.[35]

But don't start celebrating just yet. Merely living in a mixed neighbourhood isn't enough. If you seldom talk to your neighbours, diversity can actually heighten prejudice.[36] There are also indications that communities which saw a rapid influx of immigrants also had larger shares of pro-Brexit or Trump voters.[37]

Contact researchers consequently stress that people need time to get used to one another. Contact works, but not instantly. Holland witnessed fierce protests in 2015, for instance, against the opening of reception centres for Syrian refugees. Angry objectors arrived yelling and name calling, and even threw stones through windows. But then a couple of years later, that anger turned to sadness when the same asylum seekers had to be relocated elsewhere. 'We had no problems here. In fact, it was all positive,' reported one man who just a few years earlier had issued violent threats. 'It's become a place to socialise, like a community centre. I enjoy going over for a cup of coffee.'[38]

Interacting with strangers is something we have to learn, preferably starting from childhood. Best of all would be if every young person could travel like Abraham Viljoen did in his college days. Mark Twain figured that out as early as 1867, observing that 'travel is fatal to prejudice, bigotry, and narrow-mindedness'.[39]

This is not to say we need to change who we are. Quite the opposite. Among the most notable findings to come out of contact science is that prejudices can be eliminated only if we retain our own identity.[40] We need to realise it's okay that we're all different – there's nothing wrong with that. We can build strong houses for our identities, with sturdy foundations.

Then we can throw open the doors.

After visiting South Africa in 1956, Gordon Allport concluded that he'd been naive. That some societies are just too far gone, and that the weight of the past can prove too great a burden. When he died in 1967, he had no idea that all his earlier predictions would one day prove to be true.

For what had Allport asserted during one of his lectures back in Johannesburg? Yes, humans are tribal animals. Yes, we're quick to form prejudices. And yes, thinking in stereotypes seems to be rooted deep in our nature.

Yet Allport also stressed the importance of zooming out. 'To despair,' he said, 'is to misread the long lesson of history.'[41] South Africa will carry its legacy of apartheid for decades to come, but that doesn't diminish the country's breathtaking progress over the past fifty years.

Constand Viljoen passed away in April 2020. In the last years of his life, he and his brother Abraham still lived in two different worlds – one a soldier, the other a minister; one a veteran, the other a peacemaker – but the long years of not seeing each other were over. Contact had been restored.

18

When the Soldiers Came Out of the Trenches

I

On the eve of the First World War, in the summer of 1914, most thought the war would soon be over. We'll be home again by Christmas, soldiers told their sweethearts. People thronged the centres of Paris, London and Berlin, already wild with jubilation over certain triumph. Millions of recruits marched off to the front, singing as they went.

And then it began: the seminal catastrophe of the twentieth century.[1] Because if not for the First World War, there would have been no Second World War. If not for the battles of Ypres and Verdun, no Treaty of Versailles, no Russian Revolution, no Lenin, no Stalin, no Hitler.

By Christmas 1914, more than a million soldiers were dead. The front line stretched some five hundred miles, from the Belgian coast to the Franco-Swiss border. For four long years, it barely moved. Day in day out, a generation of young men was being decimated in exchange for a few acres, at most. What should have been a heroic battle, with horses, drums and trumpets, became a senseless slaughter.

But even in those desperate years, when the whole of Europe was in the grip of darkness, there was one small but radiant ray of light. In December 1914, the heavens briefly opened, giving thousands a glimpse of a different world. For a moment, they realised they were all in this together. As brothers. As humans.

It's with this story that I wish to close my book. That's because, time and again, we find ourselves back in the trenches. All too easily we forget that the other guy, a hundred yards away, is just like us. Time and again, we fire at one another from a distance – through social media or online forums, from the safety of wherever we're holed up. We let fear, ignorance, suspicion and stereotypes be our guides, making generalisations about people we've never met.

But there's an alternative. Hatred can be transformed into friendship and bitter foes can shake hands. That's something we can believe in – not because we're entitled to be naive, but because it actually happened.

2

It's Christmas Eve 1914. The night is clear and cold. Moonlight illuminates the snow-covered no man's land separating the trenches outside the town of La Chapelle-d'Armentières. British High Command, feeling nervous, sends a message to the front lines: 'It is thought possible the enemy may be contemplating an attack during Christmas or New Year. Special vigilance will be maintained during this period.'[2]

The generals have no idea what's really about to happen.

Around seven or eight in the evening, Albert Moren of the 2nd Queen's Regiment blinks in disbelief. What's that, on the other side? Lights flicker on, one by one. Lanterns, he sees, and torches, and ... Christmas trees? That's when he hears it: 'Stille Nacht, heilige Nacht'. Never before has the carol sounded so beautiful. 'I shall never forget it,' Moren says later. 'It was one of the highlights of my life.'[3]

Not to be outdone, the British soldiers start up a round of 'The First Noel'. The Germans applaud, and counter with 'O

Tannenbaum'. They go back and forth for a while, until finally the two enemy camps sing 'O Come, All Ye Faithful' in Latin, together. 'This was really a most extraordinary thing,' rifleman Graham Williams later recalled, 'two nations both singing the same carol in the middle of a war.'[4]

A Scottish regiment stationed just north of the Belgian town of Ploegsteert goes further still. From the enemy trenches, Corporal John Ferguson hears someone call out, asking if they want some tobacco. 'Make for the light,' shouts the German. So Ferguson heads out into no man's land.

'[We] were soon conversing as if we had known each other for years,' he later wrote. 'What a sight – little groups of Germans and British extending almost the length of our front! Out of the darkness we could hear laughter and see lighted matches [...] Here we were laughing and chatting to men whom only a few hours before we were trying to kill!'[5]

The next morning, Christmas Day, the bravest of the soldiers again climb out of the trenches. Walking past the barbed wire, they go over to shake hands with the enemy. Then they beckon to those who'd stayed behind. 'We all cheered,' remembered Leslie Walkington of the Queen's Westminster Rifles, 'and then we flocked out like a football crowd.'[6]

Gifts are exchanged. The British offer chocolate, tea and puddings; the Germans share cigars, sauerkraut and schnapps. They crack jokes and take group photographs as though it's a big, happy reunion. More than one game of football is played, using helmets for goalposts.[7] One match goes 3-2 to the Germans, another to the English, 4-1.

In northern France, south-west of the village of Fleurbaix, the opposing sides hold a joint burial service. 'The Germans formed up on one side,' Lieutenant Arthur Pelham-Burn later wrote, 'the English on the other, the officers standing in front, every head bared.'[8] As their comrades are laid to rest – comrades killed by enemy fire – they sing 'The Lord is my Shepherd' / 'Der Herr ist mein Hirt', their voices mingling.

That evening, there are Christmas feasts up and down the front. One English soldier finds himself escorted behind the German line to a wine cellar, where he and a Bavarian soldier pop open a bottle of 1909 Veuve Clicquot. The men exchange addresses and promise to meet up in London or Munich after the war.

You'd have a hard time believing it happened, if it weren't for all the evidence. Eyewitness accounts abound from soldiers who could scarcely believe it themselves.

'Just you think,' Oswald Tilley exclaimed to his parents in a letter, 'that while you were eating your turkey etc., I was out talking and shaking hands with the very men I had been trying to kill a few hours before!! It was astounding!'[9] German Lieutenant Kurt Zehmisch also kept having to pinch himself: 'How fantastically wonderful and strange,' he marvelled, 'that thanks to soccer and Christmas [...] deadly enemies briefly came together as friends.'[10]

Most British were stunned at how friendly the Germans were. Back home, they'd been incited by propaganda and fake news in papers such as the *Daily Mail*. More than 40 per cent of newspaper circulation was controlled by one man: Lord Northcliffe, the Rupert Murdoch of his day. He wielded tremendous power over public opinion. Germans were portrayed as ferocious Huns who went around spearing infants on bayonets and stringing priests up from church bells.[11]

Shortly before the war broke out, the German poet Ernst Lissauer had penned 'Hymn of Hate against England', which vied in popularity with the national anthem. Millions of German schoolchildren had to learn it by heart. German newspapers claimed that the French and English were so godless they didn't even celebrate Christmas.

Here, again, there was a clear pattern. The greater the distance from the front lines, the greater the hate. On the home front – in government offices and newsrooms, in living rooms and pubs – hostility towards the enemy was immense. But in the trenches, soldiers developed a mutual understanding. 'After our talk,' one British soldier wrote in a letter home, 'I really think a lot of our newspaper reports must be horribly exaggerated.'[12]

For a long time, the Christmas truce of 1914 was treated as a myth. As nothing more than a sentimental fairy tale, or, worse: a lie told by traitors. After the holidays the war resumed. Millions more soldiers were killed, and what had actually transpired that Christmas became increasingly hard to believe.

Not until the 1981 BBC documentary *Peace in No Man's Land* did it become apparent that this tale was more than just a handful of rumours. Fully two-thirds of the British front line ceased fighting that Christmas. Most instances concerned Germans who made overtures of friendship toward the British (though it also happened along Belgian and French lines). All told, more than a hundred thousand soldiers laid down their arms.[13]

In fact, the peace of Christmas 1914 was not an isolated case. The same thing happened during the Spanish Civil War and the Boer Wars. It happened in the American Civil War, in the Crimean War and in the Napoleonic Wars. But nowhere was it as widespread and sudden as that Christmas in Flanders.

Reading through the soldiers' letters, one question kept occurring to me: if these men – stuck in a horrific war that had already claimed a million lives – could come out of their trenches, then what's stopping us, here and now, from doing the same?

We, too, are being pitted against each other by hatemongers and demagogues. Newspapers like the *Daily Mail* once spread stories about bloodthirsty Huns, now they report invasions of

German soldiers celebrating Christmas in the trenches. Daily Sketch, *January 1915. Source: Getty.*

thieving foreigners, murderous immigrants and raping refugees who are – remarkably – both stealing jobs and too lazy to work, while managing to run roughshod over time-honoured traditions and values in their spare time.

This is how hate is being pumped into society once again. The culprits this time are not only newspapers, but blogs and tweets, lies spread on social media and toxic online trolls. The best fact-checker seems powerless against this kind of venom.

But what if it also works the other way around? What if propaganda not only sows discord, but can also bring people back together?

3

Colombia, 2006 – Carlos Rodriguez and Juan Pablo García work at MullenLowe, a leading global ad agency. Most days, they come up with commercials for cat food or try to sell consumers a new brand of shampoo. But on this particular day the agency gets an unusual request.

The client is Colombia's defence minister. And the job? He wants the ad agency's help in the fight against FARC, the oldest guerrilla army in Latin America. The government wants to bombard the guerrillas with guerrilla marketing.

By this time, the war in Colombia has been going on for more than fifty years and has claimed some 220,000 lives. Colombia's army, right-wing paramilitary groups, and guerrilla movements like FARC are all guilty of the most heinous war crimes. A whole generation has grown up never knowing peace. And the army realises by now that the war will never be won by brute force.

The admen at MullenLowe accept the minister's request and approach it as they would any job, by interviewing their target audience. Over the course of a year, the agency talks to nearly a hundred former FARC fighters. The researchers try to pin down what drove them into the jungle, and what keeps them there. Their conclusion after every interview is the same: these are ordinary men and women.

The rebels have the same needs, dreams and desires we all have. 'Once you really understand that they're not guerrillas, but humans,' Carlos later explained, 'the communication totally changes.'[14] In fact, the consultants arrive at exactly the same conclusions as Morris Janowitz, the psychologist who interviewed hundreds of German POWs during the Second World War (see Chapter 10). Carlos and Juan realise that instead of attacking the FARC ideology, their propaganda should aim much closer to home.

Among other things, the team discovers that the number of demobilisations peaks around the same time each year: Christmas. It seems that, like anybody else, guerrillas prefer to go home for the holidays. So the idea Carlos and Juan pitch to their boss is simple: 'Maybe we're crazy, but what would you say if we put a Christmas tree in the middle of the jungle?'[15]

Operation Christmas starts in December 2010.

Under cover of night, two special forces teams in Black Hawk helicopters fly deep into enemy territory. There, they drop two thousand Christmas lights on seventy-five-foot trees in nine strategic spots. To these 'Christmas trees' they attach motion detectors and banners that light up whenever someone walks by.

'If Christmas can come to the jungle,' it reads, 'you can come home. Demobilise. At Christmas, everything is possible.'

The operation is an overwhelming success. Within a single month, 331 guerrilla insurgents give up the fight. Many say the Christmas trees were what did it. 'Our command[er] wasn't angry,' one rebel said, 'It was different to the other propaganda we had seen ... He was touched.'[16]

Meanwhile, the team at MullenLowe continues interviewing former rebels. This is how they learn that although just about all the insurgents knew about the Christmas trees, most hadn't seen them. That's because FARC tends to travel by jungle highway – the river. And that inspires the admen's next idea.

Operation Rivers of Light launches in December 2011. Colombians who live near the rivers and have been a main source of FARC recruits are asked to write to their brothers, sisters, sons, daughters and friends who have joined the rebel army. Their message: come home, we're waiting for you.

These letters and small gifts are tucked inside 6,823 floating balls – transparent Christmas ornaments – which are then dropped into the rivers. At night, tiny lights inside the balls make the river twinkle as though lit by stars gliding into enemy territory. The result? Another 180 rebels lay down their weapons, including a FARC bomb maker.

And so it continues. The following year brings Operation Bethlehem. In the course of their interviews, Carlos and Juan learn that guerrillas often become disoriented in the jungle. Even if they wanted to go home, they can't always find their way. So the marketing agency drops thousands of little lights from military helicopters. They also set up giant beacons on the ground whose beams pierce the sky and can be seen for miles around. Rebels trying to make their way out of the jungle need only look up, like the shepherds who followed the star to Bethlehem.

Then the team decides to bring out the big guns.

If there's one thing guerrillas miss in the jungle, the guys at MullenLowe discover, it's their mothers. From the Colombian secret service they get a list of women who have children in FARC. Some haven't seen their kids for more than twenty years. Carlos and Juan ask them for old childhood snapshots of the rebels, and the team places these pictures (which only the guerrillas themselves will recognise) in parts of the jungle where FARC is fighting. The photographs all bear a simple caption: 'Before you were a guerrilla, you were my child.'

It's another hit, convincing 218 lost sons and daughters to go home to their parents.[17] Once reunited, they're granted amnesty and sent to reintegration programmes to help them learn a trade and find a job. The secret behind the whole campaign? The rebels aren't seen as monsters, but as ordinary people. 'We aren't searching for a criminal,' Juan explains, 'but for a child, missing in the jungle.'[18]

Where did all this generosity spring from? Why were the rebels offered amnesty and training and jobs? How did the people of Colombia find it in themselves to leave the past for what it was?

When I put these questions to Jose Miguel Sokoloff, Juan and Carlos' boss at MullenLowe, he laughs. 'I think our campaign may have slightly exaggerated the number of people who were willing to give the rebels a second chance.'[19]

Not that they had much choice. The agency came up against the same paradox Europe faced in 1914. The further you got from the front lines, the stronger the hatred. 'People who were never affected by the war tended to be the worst hardliners,' José confirms. But those who had actually been kidnapped themselves, or lost loved ones, wanted to put the past behind them.

The ad team decided to zero in on those stories. They decided to pretend all of Colombia would welcome returning rebels with open arms, hoping to spark off a self-fulfilling prophecy. And it worked. Thousands of guerrillas have come home since 2010, whittling down FARC's ranks in just a few years from 20,000 members to less than half that number.

Of course, this exodus wasn't all down to the operations mounted by José and his team, but at the Colombian Ministry of Defence they're convinced the peace propaganda played a crucial role. The Ministry of Finance is no doubt equally pleased with the outcome, since Christmas lights are a lot cheaper than bombs and grenades.[20]

MullenLowe's campaign provided a key impetus for the Colombian peace process that started in 2011.[21] A few years later, President Juan Manuel Santos – the defence minister who'd engaged MullenLowe – was awarded the Nobel Peace Prize. After more than half a century of fighting, the conflict was at an end. The following year, FARC handed over thousands of weapons and the last remaining fighters walked out of the jungle.

'Today is a special day,' declared President Santos. 'The day when weapons are exchanged for words.'[22]

4

This is not to say Colombia suddenly turned into some kind of peaceable kingdom. Other rebel groups still occupy the jungle, as demobilisation of the leftist FARC has now made room for far-right paramilitaries and drug traffickers. Nor will the scars from a half-century of bloodshed ever fade completely.

Even so, this story is one of hope. What the Colombian ad team witnessed was the same infectious power of kindness seen a hundred years earlier. When peace spread like an epidemic that Christmas in 1914, few soldiers were immune. One of the rare exceptions was a stiff-necked twenty-five-year-old corporal in the 16th Bavarian Reserve Infantry Regiment, who declared that 'Such things should not happen in wartime.' His name was Adolf Hitler.[23]

Most other servicemen remembered the truce in the trenches as a highlight of their lives. Again and again it was the men closest to the fighting who reached out first. From there the spirit of friendship rose through the ranks until it infected even captains, majors and colonels.

Only leaders at the very top proved resistant, as generals turned themselves inside out to halt the plague of peace. On 29 December, German Army Command issued an order that strictly prohibited fraternising with the enemy. This was echoed by a British field marshal who demanded all gestures of friendship cease.[24] Anybody who disobeyed would be court-martialled.

In subsequent years, military leaders were better prepared. At Christmas 1915, British High Command bombarded strategic positions day and night to quash any spark of Yuletide sentiment. Lieutenant Wyn Griffith of the Royal Welch Fusiliers wrote they had received 'strict orders [...] We were to remain throughout possessed by the spirit of hate, answering any advances with lead.'[25]

And yet, had it been up to many of the soldiers, the war would have ended after Christmas 1914. 'If we had been left to ourselves,' proclaimed a British major, 'there would never have been another shot fired.'[26]

Thousands of soldiers did their best to sustain the peace. Letters were passed across the lines, in secret. 'Be on guard tomorrow,' wrote a French unit to a German one. 'A general is coming to visit our position [...] we shall have to fire.' A British battalion received a similar missive from the Germans: 'We will remain your comrades. If we shall be forced to fire we will fire too high.'[27]

Soldiers at some points along the front managed to extend the ceasefire for weeks. And truces continued to break out, in spite of all the suppressive measures. When there was a mutiny among half the French divisions in 1917, the Germans didn't even realise anything was amiss. They assumed the French soldiers were merely sticking to their longstanding tacit agreement not to shoot.[28]

Throughout the war, peace threatened to erupt at any moment. Military historian Tony Ashworth describes Christmas 1914 as 'the sudden surfacing of the whole of an iceberg'.[29] For even in wartime there's a mountain of peace ready to rise up at any moment. To push that mountain back below the surface, generals, politicians and warmongers have to draw on every means at their disposal, from fake news to sheer force. Humans are simply not wired for war.

The thing we all need to remember – myself included – is that those other folks are a lot like us. The angry voter venting on TV, the refugee in the statistics, the criminal in the mugshot: every one of them is a human being of flesh and blood, someone who in a different life might have been our friend, our family, our beloved. Just like us, as one British soldier realised, 'they have people they love at home'.[30]

When we hole up in our own trenches, we lose sight of reality. We're lured into thinking that a small, hate-mongering minority reflects all humankind. Like the handful of anonymous internet trolls that are responsible for almost all the vitriol on Twitter and Facebook.[31] And even the most caustic keyboard crusader may at other times be a thoughtful friend or loving caregiver.

To believe people are hardwired to be kind isn't sentimental or naive. On the contrary, it's courageous and realistic to believe in peace and forgiveness. Jose Miguel Sokoloff tells of an officer in one of the Colombian army units that helped spread the ad agency's Christmas message. A few months later, the officer was killed in action. Jose still gets emotional remembering what he learned from his friend. 'I want to do this,' the officer told him, 'because generosity makes me stronger. And it makes my men feel stronger, too.'[32]

That's a truth as old as time. Because, like all the best things in life, the more you give, the more you have. That's true of trust and friendship, and it's true of peace.

Epilogue

TEN RULES TO LIVE BY

'If you make a film about a man kidnapping a woman and chaining her to a radiator for five years – something that has happened probably once in history – it's called searingly realistic analysis of society. If I make a film like *Love Actually*, which is about people falling in love, and there are about a million people falling in love in Britain today, it's called a sentimental presentation of an unrealistic world.'

Richard Curtis

Legend has it that two words were inscribed in the forecourt of the Temple of Apollo in Delphi. The temple was a major pilgrimage site, and visitors came here from all corners of ancient Greece in search of divine counsel.

What they read, upon entering, was: *GNOTHI SEAUTON*. Know thyself.

Considering the most recent evidence of psychology and biology, of archaeology and anthropology, of sociology and history, we can only conclude that humans have for millennia navigated by a faulty self-image. For ages, we've assumed that people are selfish, that we're beasts, or worse. For ages, we've believed civilisation is a flimsy veneer that will crack at the merest provocation. Now we know this view of humankind, and this perspective on our history, is utterly unrealistic.

In the last chapters of this book I've attempted to present the new world that awaits if we revise our view of human nature. I've probably only scratched the surface. After all, if we believe most people are decent and kind, everything changes. We can completely rethink how we organise our schools and prisons, our businesses and democracies. And how we live our own lives.

At this point, I should point out that I'm not a fan of the self-help genre. If you ask me, we're living in an age of too much introspection and too little outrospection. A better world doesn't begin with me, but with all of us, and our main task is to build different institutions. Another hundred tips for climbing the career ladder or visualising your way to wealth won't get us anywhere.

But then a friend asked me if writing this book had changed my own view on life, and I realised that the answer is yes. A realistic view of human nature can't help but have major implications for how you interact with other people. So for what it's worth, here are my own ten rules to live by, based on what I've learned over the past few years.

I: When in doubt, assume the best

My first commandment for myself is also the hardest. In Chapter 3 we saw that humans have evolved to connect, but communication can be tricky. You say something that gets taken the wrong way, or someone looks at you funny, or nasty comments get passed through the grapevine. In every relationship, even one built on years of matrimony, we often don't know what the other person is thinking about us.

And so we guess. Say I suspect a co-worker doesn't like me. Regardless of whether that's true, my behaviour is certain to alter in ways that won't help our relationship. In Chapter 1 we saw that people have a *negativity bias*. A single unpleasant remark makes a deeper impression than ten compliments combined (the bad may seem stronger, but it's outnumbered by the good.). And when in doubt, we're inclined to assume the worst.

Meanwhile, we fall victim to what's known as *asymmetrical feedback*. Basically, this means that if your faith in someone is misplaced, the truth will surface sooner or later. You'll discover that your best friend has fled the country with your life savings, or the deal on that fixer-upper was indeed too good to be true, or after six weeks of using the Ab King Pro you still don't have the six pack promised on TV. If you've been too trusting, eventually you find out.[1]

But if you decide *not* to trust someone, you'll never know if you're right. Because you'll never get any feedback. Let's say you get screwed over by some blond Dutch guy, so you swear never again to trust blond people from Holland. For the rest of your life, you'll be suspicious of *all* blond Dutchies, without ever having to face the simple truth that most of them are pretty decent.

So when in doubt about another person's intentions, what should you do?

It's most realistic to assume the best – to give them the benefit of the doubt. Usually this is justified, because most people mean well. And in the rare case that someone does try to pull one over on you, how you respond could well have a non-complementary effect.[2] (Think back to Julio Diaz, who took his would-be mugger out to dinner.)

But what if you still get scammed? Psychologist Maria Konnikova talks about this in her fascinating book on professional con artists.[3] You might expect her main tip would be to always be on guard. But no. Konnikova, the leading expert on frauds and swindles, comes to a very different conclusion. Far better, she says, is to accept and account for the fact that you'll occasionally be cheated. That's a small price to pay for the luxury of a lifetime of trusting other people.

Most of us feel ashamed when our faith turns out to have been misplaced. But maybe, if you're a realist, you ought to feel a little proud. In fact, I'd go even farther: if you've never been conned, then you should be asking if your basic attitude is trusting enough.

II: Think in win-win scenarios

The story goes that Thomas Hobbes was strolling around London with a friend one day when he stopped suddenly to give a beggar some money. His friend was surprised. Hadn't Hobbes himself said that it's in our nature to be selfish? The philosopher didn't see a problem. Witnessing the beggar's suffering caused Hobbes discomfort, so it felt good to give the man a few coins. Ergo, his action was motivated by self-interest.[4]

For the last couple of centuries, philosophers and psychologists have racked their brains over the question whether there is such a thing as pure selflessness. But to be honest, that whole debate doesn't really interest me. Because just imagine living in a world where you got a sick feeling every time you performed a kind act. What sort of hell would that be?

The wonderful fact is that we live in a world where doing good also feels good. We like food because without food we'd starve. We like sex because without sex we'd go extinct. We like helping because without each other we'd wither away. Doing good typically feels good because it *is* good.

Sadly, untold companies, schools and other institutions are still organised around a myth: that it's in our nature to be locked in competition with one another. 'In a great deal you win – not the other side,' counsels Donald Trump in his book *Think Big and Kick Ass*. 'You crush the opponent and come away with something better for yourself.'[5]

In truth, this works precisely the other way around. The best deals are those where *everybody* wins. Those prisons in Norway? They're better, more humane and less expensive. Jos de Blok's homecare organisation in Holland? It delivers higher quality at lower cost, pays employees more and leaves both staff and patients more satisfied. These are scenarios where everybody wins.

In the same vein, the literature on forgiveness emphasises that forgiving others works in our own self-interest.[6] It's not only a gift, but a good deal, because to forgive is to stop wasting your energy on antipathy and grudges. Effectively, you liberate yourself to live. 'To forgive is to set a prisoner free,' wrote the theologian Lewis B. Smedes, 'and discover that the prisoner was you.'[7]

III: Ask more questions

The Golden Rule of virtually every philosophy in world history is some form of: 'Do not do unto others as you would not have them do unto you.' This bit of wisdom was already being expounded by the Chinese thinker Confucius two and a half thousand years ago. It turns up again with the Greek historian Herodotus and the philosophy of Plato, and a few centuries later the rule was encoded in Jewish, Christian and Islamic scriptures.

These days, billions of parents repeat the Golden Rule to their children. It comes in two flavours: the positive injunction ('Treat others as you wish to be treated') and the negative injunction ('Do *not* do unto others what you would not have them …'). Some neurologists even believe the rule is a product of millions of years of human evolution and is programmed into our brains.[8]

Even so, I've come to believe the Golden Rule falls short. In Chapter 10, we saw that empathy can be a bad guide: the simple fact is we're not always good at sensing what others want. All those managers, CEOs, journalists and policymakers who think they do are effectively robbing others of their voice. This is why you so seldom see refugees interviewed on TV. This is why our democracy and journalism constitute mostly one-way traffic. And this is why our welfare states are steeped in paternalism.

Far better would be to start by asking a question. To let citizens have their say, as in the participatory democracy in Porto Alegre (see Chapter 15). To let employees direct their own teams, as in Jean-François Zobrist's factory (see Chapter 13). To let kids plot their own learning pathways, as in Sjef Drummen's school (see Chapter 14).

This variation on the familiar maxim, also known as the 'Platinum Rule', was nicely summed up by George Bernard Shaw. 'Do *not* do unto others as you would that they should do unto you,' he advised. 'Their tastes may be different.'[9]

IV: Temper your empathy, train your compassion

The Platinum Rule calls not for empathy, but compassion. To help explain the difference, let me introduce the Buddhist monk Matthieu Ricard, a man with a legendary command of his thoughts. (If this appeals to you, I can only say good luck meditating for the 50,000 hours it took him to get there.)

Not long ago, Ricard was invited by neurologist Tania Singer to spend a morning in her brain scanner.[10] Singer wanted to know what happens in our brains when we feel empathy. More important, she wanted to know if there's an alternative.

To prepare, Singer had Ricard watch a documentary the evening before about lonely orphans in a Romanian institution. When sliding his brain under the scanner the next day, Singer asked him to recall their vacant eyes. Their spindly limbs. Ricard did as she asked, imagining as fiercely as he could how those Romanian orphans felt.

An hour later, he was a wreck.

Because that's what empathy does to us. It's exhausting. In a later experiment, Singer asked a group of volunteers to spend fifteen minutes with their eyes shut evoking as much empathy as possible, every day for one week. This was about as long as they could endure. At the end of the week, all participants were more pessimistic. One woman said that when she looked at fellow passengers in the train afterwards, all she could see was suffering.[11]

After her first session with Ricard, Singer decided to try something different. Once again, she asked the monk to think about the Romanian orphans, but this time he wasn't to imagine himself in their shoes. Rather, she wanted him to apply the skill he'd spent years perfecting, feeling not *with* them, but *for* them. Instead of sharing in their distress, Ricard concentrated on calling up feelings of warmth, concern and care. Instead of personally experiencing their suffering, he kept himself removed from it.

On her monitor, Singer could instantly see the difference, because wholly different parts of Ricard's brain lit up. Empathy mostly activates the *anterior insula*, which sits just above our ears, but flashing now were his *corpus striatum* and *orbitofrontal cortex*.

What was going on? Ricard's new mentality is what we call compassion. And, unlike empathy, compassion doesn't sap our energy. In fact, afterwards Ricard felt much better. That's because compassion is simultaneously more controlled, remote and constructive. It's not about sharing another person's distress, but it does help you to recognise it and then act. Not only that, compassion injects us with energy, which is exactly what's needed to help.

To give another example, let's say your child is afraid of the dark. As a parent, you're not going to crouch in a corner of the room and whimper alongside your son or daughter (empathy). Rather, you try to calm and comfort them (compassion).

So should we all start meditating like Matthieu Ricard? I confess it sounded a little new-agey to me at first, but there is some scientific evidence that meditation can train our compassion.[12] The brain is a malleable organ. And if we exercise to keep our bodies in shape, why not do the same for our minds?

V: Try to understand the other, even if you don't get where they're coming from

To be honest, I gave meditating a shot, but it hasn't been a huge success so far. For some reason there's always another email, another tweet or another video of a goat on a trampoline demanding immediate attention. And meditating for 50,000 hours? Sorry, but I also have a life.

Lucky for me, there's another way to zoom out: use the method of choice among Enlightenment philosophers back in the eighteenth century. What is it? Reason. Intellect. Our capacity to put things into rational perspective is a psychological process that enlists different parts of our brain. When we use our intellect to try to understand someone, this activates the *prefrontal cortex*, an area located just behind the forehead that's exceptionally large in humans.[13]

Sure, I know there are scores of studies on the thousand and one cases in which this cortex drops the ball. Studies reveal we're often not so rational and self-possessed after all. Nonetheless, I think it's important not to overstate such findings. We use rational arguments and evidence all the time in day-to-day life, and we've built societies full of laws and rules and agreements. Humans think much better than we think. And our powers of reason are not a thin coating covering our emotional nature, but an essential feature of who we are and what makes us human.[14]

Take Norway's vision on prisons, which can seem counter-intuitive to the rest of us. By applying our intellect and examining recidivism statistics, we realise it's an excellent way to deal with criminals. Or take Nelson Mandela's ethic of statesmanship. Over and over he had to bite his tongue, tamp down on his emotions and stay sharp and analytical. Mandela was not only kind-hearted, he was equally astute. Having faith in others is as much a rational decision as an emotional one.

Of course, seeing where someone else is coming from doesn't mean you need to see eye to eye. You can understand the mindset of a fascist, a terrorist, or a fan of *Love Actually* without jumping on the fascist, terrorist, or lover-of-sappy-movies bandwagon. (I have to say, I'm a proud member of that last group.) Understanding the other at a rational level is a skill. It's a muscle you can train.

Where we need our capacity for reason most of all is to *suppress*, from time to time, our desire to be nice. Sometimes our sociable instinct gets in the way of truth and of equity. Because consider: haven't we all seen someone treated unfairly yet kept silent to avoid being disagreeable? Haven't we all swallowed our words just to keep the peace? Haven't we all accused those who fight for their rights of rocking the boat?

I think that's the great paradox of this book. I've argued that humans have evolved to be fundamentally sociable creatures, but sometimes our sociability is the problem. History teaches that progress often begins with people — like Buurtzorg's Jos de Blok and Agora's Sjef Drummen — whom others feel to be preachy or even *unfriendly*. People with the nerve to get on their soapbox at social occasions. Who raise unpleasant subjects that make you uneasy.

Cherish these people, because they're the key to progress.

VI: Love your own as others love their own

On 17 July 2014, a Malaysia Airlines Boeing 777 went down outside the village of Hrabove in Ukraine. On board were 298 passengers, 193 of them Dutch. The aircraft had been shot down by pro-Russian separatists. Nobody survived the crash.

At first, the reports — those 298 deaths — felt abstract, but then I read a story in a Dutch newspaper that struck me in the gut.[15]

It opened with a picture of Karlijn Keijzer (aged twenty-five) and Laurens van der Graaff (thirty); a selfie of the beaming faces of a blond guy and a curly haired girl captured just before they boarded the plane. I read that they'd met at an Amsterdam rowing club. That Laurens wrote for *Propria Cures*, a brilliant student newspaper, and that Karlijn had almost finished her Ph.D. in the States.

And that they were crazy about each other.

'They will always be that head-over-heels-in-love, unable-to-keep-their-hands-off-each-other happy couple,' a friend was quoted as saying. Isn't it hypocritical, I asked myself, that having just skipped over a piece on atrocities in Iraq on page seven, now I was tearing up? Normally this kind of reporting bothers me. 'Two Dutch citizens dead off the coast of Nigeria,' the papers report, when a whole boatload of people went down.

But humans are limited creatures. We care more about those who are like us, who share the same language or appearance or background. I too was a Dutch college student once and joined a university club. I too met a girl there with gorgeous curls and I would have loved to write for *Propria Cures*. ('For those who knew Laurens,' his colleagues there wrote, 'it came as no surprise that it would take an anti-aircraft missile to stop his powerful body.')[16]

It was Karlijn's brother who sent the paper that smiling selfie taken hours before their deaths. 'The only thing I ask,' he wrote, 'is that you show the country and the world the pain that I and my other sister and parents are going through. This is the pain of hundreds of people in the Netherlands.'

And he was right. Everyone knew somebody who'd known somebody on that plane. During those days, I felt Dutch in a way I never had before.

Why do we care more about people who seem like us? In Chapter 10, I wrote that evil does its work from a distance.

Distance lets us rant at strangers on the internet. Distance helps soldiers bypass their aversion to violence. And distance has enabled the most horrifying crimes in history, from slavery to the Holocaust.

But choose the path of compassion and you realise how little separates you from that stranger. Compassion takes you beyond yourself, until those near and dear are no more or less significant than the rest of the world. Why else did the Buddha abandon his family? Why else would Jesus have instructed his disciples to leave behind their fathers and mothers, their wives and children, their brothers and sisters?

But maybe you can take this too far.

Maybe love of one's fellow man starts small. If a person is filled with self-loathing, how can they possibly love anyone else? If someone loses sight of family and friends, how can they shoulder the burdens of this world? We can't take on the big until we have a handle on the small. Among those 193 Dutch passengers were many men and women who were unabashedly making the world a better place, from AIDS researchers to human rights advocates. And yet, the greatest loss was to the people who were closest to them.

As humans, we differentiate. We play favourites and care more about our own. That's nothing to be ashamed of – it makes us human. But we must also understand that those others, those distant strangers, also have families they love. That they are every bit as human as we are.

VII: Avoid the news

One of the biggest sources of distance these days is the news. Watching the evening news may leave you feeling more attuned to reality, but the truth is that it skews your view of the world.

The news tends to generalise people into groups like politicians, elites, racists and refugees. Worse, the news zooms in on the bad apples.

The same is true of social media. What starts as a couple of bullies spewing hate speech at a distance gets pushed by algorithms to the top of our Facebook and Twitter feeds. It's by tapping into our negativity bias that these digital platforms make their money, turning higher profits the worse people behave. Because bad behaviour grabs our attention, it's what generates the most clicks, and where we click the advertising dollars follow.[17] This has turned social media into systems that amplify our worst qualities.

Neurologists point out that our appetite for news and push notifications manifests all the symptoms of addiction, and Silicon Valley figured this out long ago. Managers at companies like Facebook and Google strictly limit the time their children spend on the internet and 'social' media. Even as education gurus sing the praises of iPads in schools and digital skills, the tech elites, like drug lords, shield their own kids from their toxic enterprise.[18]

My rule of thumb? I have several: steer clear of television news and push notifications and instead read a more nuanced Sunday paper and in-depth feature writing, whether online or off. Disengage from your screen and meet real people in the flesh. Think as carefully about what information you feed your mind as you do about the food you feed your body.

VIII: Don't punch Nazis

If you're an avid follower of the news, it's easy to get trapped by hopelessness. What's the point of recycling, paying taxes and donating to charities when others shirk their duty?

If you're tempted by such thoughts, remember that cynicism is just another word for laziness. It's an excuse not to take responsibility. Because if you believe most people are rotten, you don't need to get worked up about injustice. The world is going to hell either way.

There's also a kind of activism that looks suspiciously like cynicism. This is the do-gooder who's mostly concerned with their own self-image. To go down this path is to become the rebel who knows best, doling out advice without any genuine regard for others. Bad news is then good news, because bad news ('Global warming is speeding up!', 'Inequality is worse than we thought!') proves they were right all along.[19]

But there's a different way, as the small town of Wunsiedel in Germany shows. In the late 1980s, Adolf Hitler's deputy Rudolf Hess was buried in the local cemetery, and Wunsiedel rapidly became a neo-Nazi pilgrimage site. Even today, skinheads march through town every year on 17 August, the anniversary of Hess's death, hoping to incite riots and violence.

And every year, right on cue, anti-fascists come along to give the neo-Nazis exactly what they want. Inevitably, a video surfaces showing someone proudly taking a swing at some Nazi. But afterwards, the effects prove counter-productive. Just like bombing the Middle East is manna for terrorists, punching Nazis only reinforces extremists. It validates them in their worldview and makes it that much easier to attract new recruits.

Wunsiedel decided to test a different strategy. In 2014 a wisecracking German named Fabian Wichmann had a brilliant idea. What if the town turned the march for Rudolf Hess into a charity walk? Residents loved the idea. For every metre the neo-Nazis walked, the townspeople pledged to donate ten euros to Wichmann's organisation EXIT-Deutschland, which helps people get out of far-right groups.

Ahead of the event, the townspeople marked off start and finish lines. They made banners thanking the walkers for their efforts. The neo-Nazis, meanwhile, had no idea what was afoot. On the day itself, Wunsiedel greeted them with loud cheers and showered them with confetti upon crossing the finish. All told, the event raised more than twenty thousand euros for the cause.

Wichmann emphasises that the important thing after a campaign like this is to keep the door open. In the summer of 2011 his organisation handed out T-shirts at an extremist rock festival in Germany. Emblazoned with far-right symbols, the shirts at first seemed to endorse neo-Nazi ideology. But after being washed, a different message appeared. 'What your T-shirt can do, you can do too. We can help you free yourself from the far-right.'[20]

This may sound cheesy, but in subsequent weeks the number of phone calls to EXIT-Deutschland went up 300 per cent. Wichmann saw how disorienting his messages were to the neo-Nazis. Where they'd expected disgust and outrage, they got an outstretched hand.

IX. Come out of the closet: don't be ashamed to do good

To extend that hand you need one thing above all. Courage. Because you may well be branded a bleeding heart or a show-off. 'When you give to the needy, sound no trumpet ...' Jesus warned during the Sermon on the Mount, and 'when you pray, go into your room and shut the door and pray to your Father who is in secret'.[21]

On the face of it, this sounds like sensible advice. Who wants to be thought sanctimonious? Much safer to do your good deeds in secret, or at the very least be prepared with an excuse:

'Just keeping busy.'

'I didn't need the money anyway.'

'It'll look good on my résumé.'

Modern psychologists have discovered that when people do something out of the goodness of their hearts, they often fabricate selfish motives. This is most prevalent in individualistic Western cultures where veneer theory is most entrenched.[22] And it makes sense: if you assume most people are selfish, then any good deed is inherently suspect. As one American psychologist notes, 'People seem loathe to acknowledge that their behaviour may have been motivated by genuine compassion or kindness.'[23]

Unfortunately, this reticence works like a nocebo. When you disguise yourself as an egotist, you reinforce other people's cynical assumptions about human nature. Worse, by cloaking your good deeds, you place them in quarantine, where they can't serve as an example for others. And that's a shame, because *Homo puppy*'s secret superpower is that we're so great at copying one another.

Don't get me wrong: inspiring others is not about flaunting your deeds, and championing the good doesn't mean blowing your own trumpet. In the Sermon on the Mount, Jesus cautioned his disciples against the one, while he encouraged the other: 'You are the light of the world. A city set on a hill cannot be hidden. Nor do people light a lamp and put it under a basket, but on a stand, and it gives light to all in the house. In the same way, let your light shine before others, so that they may see your good works ...'[24]

That doing good can be contagious was demonstrated by two American psychologists with a brilliant experiment in 2010.[25] They set up a game for stakes in which 120 volunteers, all strangers to each other, were split up into four-player groups. Each person got some cash to start off with and then was free to choose if and how much to contribute to the group kitty. After the first round all the groups were shuffled so that no two people were ever in the same one twice.

What happened next was a veritable multiplying money trick. Whenever someone contributed an additional dollar to the kitty in the first round, other players in the group contributed on average twenty cents more in the next round, *even though they were playing with different people*. This effect persisted into the third round, where players contributed an average of five cents more. In the final tally, every contribution of one dollar was more than doubled.

I think back to this study often because this is something I want to keep in mind. Every good deed is like a pebble in a pond, sending ripples out in all directions. 'We don't typically see how our generosity cascades through the social network,' noted one of the researchers, 'to affect the lives of dozens or maybe hundreds of other people.'[26]

Kindness is catching. And it's so contagious that it even infects people who merely see it from afar. Among the first psychologists to study this effect was Jonathan Haidt, in the late 1990s.[27] In one of his articles he tells the story of a student who helped an old lady shovel snow from her driveway. One of his friends, seeing this selfless act, later wrote: 'I felt like jumping out of the car and hugging this guy. I felt like singing and running, or skipping and laughing. Just being active. I felt like saying nice things about people. Writing a beautiful poem or love song. Playing in the snow like a child. Telling everybody about his deed.'[28]

Haidt discovered that people are often surprised and moved by simple acts of generosity. When the psychologist asked his research subjects how this kind of experience affected them, they described an irresistible urge to go out and help someone, too.

Haidt calls this emotion 'elevation'. People are wired so that a simple sign of kindness literally makes us feel warm and tingly.

And what's fascinating is that this effect occurs even when we hear these stories from someone else. It's as though we press a mental reset button that wipes away our cynical feelings so we once more have a clear view of the world.

X. Be realistic

And now for my most important rule to live by.

If there's one thing I've sought to do with this book, it's to change the meaning of the word 'realism'. Isn't it telling that in modern usage the realist has become synonymous with the cynic – for someone with a pessimistic outlook?

In truth, it's the cynic who's out of touch. In truth, we're living on Planet A, where people are deeply inclined to be good to one another.

So be realistic. Be courageous. Be true to your nature and offer your trust. Do good in broad daylight, and don't be ashamed of your generosity. You may be dismissed as gullible and naive at first. But remember, what's naive today may be common sense tomorrow.

It's time for a new realism. It's time for a new view of humankind.

ACKNOWLEDGEMENTS

In January 2013 I got a message from Dutch philosopher Rob Wijnberg asking if I wanted to grab coffee. He told me he wanted to discuss plans for launching a new journalism platform. He envisaged a publication with no news, no advertising and no cynicism. Instead, we would offer solutions.

Within months, what would become *De Correspondent* had set a new world record in crowdfunding, and I had a new job. This book is the result of seven years of working at *De Correspondent*. It's the product of innumerable conversations with readers who honed, improved, or overturned my ideas. And it's the result of the privilege of being able to pursue my own fascinations and be powered by that magical stuff known as intrinsic motivation.

My thanks go to all my colleagues there. To Rob, of course, who energises me like no one else. To Jesse Frederik, who taught me to be more critical of my own ideas. To Milou Klein Lankhorst, who once more proved the best book publisher in Europe. And to Andreas Jonkers, whose contributions as deputy book publisher were invaluable for this book.

I had the good fortune that Harminke Medendorp agreed to edit the original Dutch text. Harminke is among the best in her field, able with a few pointed questions to make you realise what you're actually trying to say. My thanks also to all the colleagues who read versions of the Dutch manuscript: Tomas Vanheste, Maurits Martijn, Rosan Smits, Marnix de Bruyne, Sanne Blauw,

Michiel de Hoog, Johannes Visser, Tamar Stelling, Jelmer Mommers, Arjen van Veelen, Maite Vermeulen, Riffy Bol, Charlotte Remarque, and Anna Vossers. When you get to work with people like them, it's hard to be cynical.

I would also like to thank Matthias van Klaveren, Sem de Maagt, Huib ter Horst and Carlijn Kingma, who read parts of the original book and offered valuable advice. Carlijn is one of the most talented artists in Europe and the artwork she created based on this book will probably hang in galleries long after my pages have all been recycled.

For this English translation, I'm immensely grateful to Elizabeth Manton and Erica Moore. Translation is a difficult and often undervalued craft, and an art they've mastered like no other. I also thank my editors Ben George at Little, Brown and Alexis Kirschbaum at Bloomsbury, who helped further hone the text, my literary agents Rebecca Carter and Emma Parry, who believed in this book from the start, and copy editor Richard Collins, for his excellent work.

Finally, I owe a huge debt to my family, to my sisters and brothers-in-law and to my friends. To Jurriën, for being a wonderful friend. To Maartje, for everything (including the English book title). And to my parents, Peta and Kees Bregman, to whom this book is dedicated.

Prologue

1   Churchill said this in the House of Commons on 30 July 1934.
2   J. F. C. Fuller, *The Reformation of War* (London, 1923), p. 150.
3   Gustave Le Bon, *The Crowd. A Study of the Popular Mind* (Kitchener, 2001), p. 19. Originally published in 1896.
4   Richard Overy, 'Hitler and Air Strategy', *Journal of Contemporary History* (July 1980), p. 410.
5   J. T. MacCurdy, *The Structure of Morale* (Cambridge, 1943), p. 16.
6   Quoted in Richard Overy, *The Bombing War. Europe 1939–1945* (London, 2013), p. 185.
7   Angus Calder, *The People's War. Britain 1939–1945* (London, 1991), p. 174.
8   Overy, *The Bombing War*, p. 160.
9   Robert Mackay, *Half the Battle: Civilian Morale in Britain During the Second World War* (Manchester, 2002), p. 261.
10 Quoted in Overy, *The Bombing War*, p. 145. In early 1941, only 8 per cent of bomb shelters were still being used. See Overy, p. 137.
11 Sebastian Junger, *Tribe. On Homecoming and Belonging* (London, 2016).
12 Richard Overy, 'Civilians on the frontline', *Observer* (6 September 2009).
13 Mollie Panter-Downes, *London War Notes 1939–1945* (New York, 1971), p. 105.
14 Overy, *The Bombing War*, p. 264.
15 Even friends who knew Frederick Lindemann well characterised him as someone who 'always thought that he was in the right about everything, and was never prepared to yield or admit failure', 'was prone to regard views opposite to his own as a personal insult' and 'was never deterred from pontificating about a subject because he did not understand it'. See Hugh Berrington, 'When does Personality

Make a Difference? Lord Cherwell and the Area Bombing of Germany', *International Political Science Review* (January 1989).

16 Quoted in Brenda Swann and Francis Aprahamian, *J. D. Bernal. A Life in Science and Politics* (London, 1999), p. 176. Two thousand children were assigned to write essays about their experiences. Reading those essays today, their bravery is astonishing. 'I was buried, I was cut but I still helped to pull out the dead and injured,' one ten-year-old boy wrote of his destroyed house. See Martin L. Levitt, 'The Psychology of Children: Twisting the Hull-Birmingham Survey to Influence British Aerial Strategy in World War II', *Psychologie und Geschichte* (May 1995).

17 Solly Zuckerman, *From Apes to Warlords. An Autobiography, 1904–1946* (London, 1988), p. 405. In the first edition of this book, published in 1978, Zuckerman added the title page of the Hull report as an appendix, thus violating the embargo that was in force until 2020.

18 Quoted in Charles Webster and Noble Frankland, *The Strategic Air Offensive Against Germany 1935–1945* (London, 1961), p. 332.

19 C. P. Snow, 'Whether we live or die', *Life* magazine (3 February 1961), p. 98.

20 Overy, *The Bombing War*, p. 356.

21 Quoted in Jörg Friedrich, *The Fire. The Bombing of Germany 1940–1945* (New York, 2006), p. 438.

22 Quoted in Friedrich Panse, *Angst und Schreck* (Stuttgart, 1952), p. 12.

23 Friedrich, *The Fire*, pp. 418–20.

24 The British report was not released until fifty years later. See Sebastian Cox (ed.), *British Bombing Survey Unit, The Strategic Air War Against Germany, 1939–1945. The Official Report of the British Bombing Survey Unit* (London, 1998).

25 John Kenneth Galbraith, *A Life in Our Times* (Boston, 1981), p. 206. The million-dollar question, of course, is what would have happened if the Allies had invested less in their air force and more in the army and the navy? After the Second World War, Nobel laureate Patrick Blackett wrote that the war would have ended six to twelve months earlier. And the Germans reached the same conclusion. Albert Speer, minister of armaments and war production, said he had been most concerned by attacks on German infrastructure, while *Luftwaffe* commander Hermann Göring mainly remembered the strikes on the German oil refineries. By the fall of 1944, German oil reserves were dwindling. Tanks stalled, aircraft were stuck in their hangars and artillery were being carted around by horses. But that didn't deter

the British from bombing German civilians. In the last three months
of 1944, 53 per cent of the bombings targeted urban areas, and only
14 per cent were aimed at oil refineries. By that time the British had
virtually stopped using firebombs, knowing there was not much left
to burn. In the meantime, German oil production resumed. See Max
Hastings, *Bomber Command* (London, 1979), pp. 327–34.

26 Edward Miguel and Gerard Roland, 'The Long Run Impact of
Bombing Vietnam', *Journal of Development Economics* (September
2011), p. 2.

1 A New Realism

1 Tom Postmes, email to the author, 9 December 2016.

2 Jack Winocour (ed.), *The Story of the Titanic As Told by Its Survivors*
(New York, 1960), p. 33.

3 Quoted in Rebecca Solnit, *A Paradise Built in Hell. The Extraordinary
Communities that Arise in Disaster* (New York, 2009), p. 187.

4 Frans de Waal, *The Bonobo and the Atheist. In Search of Humanism
Among the Primates* (New York, 2013), p. 43.

5 Gary Younge, 'Murder and Rape – Fact or Fiction?', *Guardian* (6
September 2005).

6 Quoted in Robert Tanner, 'New Orleans Mayor Orders Police Back
to Streets Amid Increasingly Violent Looting', *Seattle Times* (1
September 2005).

7 Timothy Garton Ash, 'It Always Lies Below', *Guardian* (8 September
2005).

8 Jim Dwyer and Christopher Drew, 'Fear Exceeded Crime's Reality
in New Orleans', *New York Times* (29 September 2005).

9 Havidán Rodríguez, Joseph Trainor and Enrico L. Quarantelli,
'Rising to the Challenges of a Catastrophe: The Emergent and
Prosocial Behavior Following Hurricane Katrina', *The Annals of the
American Academy of Political and Social Science* (No. 1, 2006).

10 Matthieu Ricard, *Altruism. The Power of Compassion to Change
Yourself and the World* (New York, 2015), p. 99.

11 Enrico L. Quarantelli, 'Conventional Beliefs and Counterintuitive
Realities', *Social Research: An International Quarterly of the Social
Sciences* (No. 3, 2008), p. 885.

12 Quoted in AFP/Reuters, 'Troops Told "Shoot to Kill" in New
Orleans' (2 September 2005).

13 Trymaine Lee, 'Rumor to Fact in Tales of Post-Katrina Violence',
    New York Times (26 August 2010).

14 Solnit, *A Paradise Built in Hell*, p. 131.

15 Quoted in CNN Money, 'Coke Products Recalled' (15 June 1999).

16 B. Nemery, B. Fischler, M. Boogaerts, D. Lison and J. Willems,
    'The Coca-Cola Incident in Belgium, June 1999', *Food and Chemical
    Toxicology* (No. 11, 2002).

17 Victoria Johnson and Spero C. Peppas, 'Crisis Management in
    Belgium: the case of Coca-Cola', *Corporate Communications: An
    International Journal* (No. 1, 2003).

18 Quoted in Bart Dobbelaere, 'Colacrisis was massahysterie', *De
    Standaard* (2 April 2000).

19 Karolina Wartolowska et al., 'Use of Placebo Controls in the
    Evaluation of Surgery: Systematic Review', *British Medical Journal*
    (21 May 2014).

20 Clayton R. Critcher and David Dunning, 'No Good Deed Goes
    Unquestioned: Cynical Reconstruals Maintain Belief in the Power
    of Self-interest', *Journal of Experimental Social Psychology* (No. 6,
    2011), p. 1212.

21 Sören Holmberg and Bo Rothstein, 'Trusting other people', *Journal
    of Public Affairs* (30 December 2016).

22 Jodie Jackson, 'Publishing the Positive. Exploring the Motivations
    for and the Consequences of Reading Solutions-focused Journalism',
    constructivejournalism.org (Fall 2016).

23 See, for example, Wendy M. Johnston and Graham C. L. Davey,
    'The psychological impact of negative TV news bulletins: The
    catastrophizing of personal worries', *British Journal of Psychology* (13
    April 2011).

24 Hans Rosling, *Factfulness* (London, 2018), p. 50.

25 Chris Weller, 'A top economist just put the fight against poverty in
    stunning perspective', *Business Insider* (17 October 2017).

26 Toni van der Meer et al., 'Mediatization and the Disproportionate
    Attention to Negative News. The case of airplane crashes',
    Journalism Studies (16 January 2018).

27 Laura Jacobs et al., 'Back to Reality: The Complex Relationship
    Between Patterns in Immigration News Coverage and Real-World
    Developments in Dutch and Flemish Newspapers (1999–2015)',
    Mass Communication and Society (20 March 2018).

28 Nic Newman (ed.), *Reuters Institute Digital News Report. Tracking
    the Future of News* (2012). See also Rob Wijnberg, 'The problem

with real news – and what we can do about it', *Medium.com* (12
September 2018).

29 Quoted in Michael Bond, 'How to keep your head in scary
situations', *New Scientist* (27 August 2008).

30 Rolf Dobelli, 'Avoid News. Towards a Healthy News Diet', *dobelli.
com* (August 2010).

31 Frans de Waal, *The Bonobo and the Atheist*, pp. 38–9.

32 Michael Ghiselin, *The Economy of Nature and the Evolution of Sex*
(Berkeley, 1974), p. 247.

33 Joseph Henrich et al., 'In Search of Homo Economicus: Behavioral
Experiments in 15 Small-Scale Societies', *American Economic Review*
(No. 2, 2001).

34 David Sloan Wilson and Joseph Henrich, 'Scientists Discover What
Economists Haven't Found: Humans', *Evonomics.com* (12 July 2016).

35 Quoted in David Sloan Wilson, 'Charles Darwin as the Father of
Economics: A Conversation with Robert Frank', *The Evolution
Institute* (10 September 2015).

36 Thucydides, *History of the Peloponnesian War*, translated by Rex
Warner (1972), pp. 242–5.

37 Saint Augustine, *The Confessions of Saint Augustine*, translated by
Maria Boulding (2012), p. 12.

38 Thomas Henry Huxley, *The Struggle for Existence in Human Society*
(originally published in 1888).

39 Herbert Spencer, *Social Statistics*, Chapter XVIII, paragraph 4 (1851).

40 'I refuse to believe that the tendency of human nature is always
downward,' said Mahatma Gandhi, the legendary leader of India's
independence movement, whom Churchill dismissed as a 'half-
naked fakir'. 'Man's goodness is a flame that can be hidden but never
extinguished,' said Nelson Mandela, who was imprisoned for twenty-
seven years by a criminal regime.

41 Emma Goldman, *Anarchism and Other Essays* (Stillwell, 2008), p. 29.
Originally published in 1910.

42 This was Marie Lindegaard, whom we'll meet in Chapter 9.

2 The Real *Lord of the Flies*

1 William Golding recalled this in the introduction to his audiotaped
reading of the book produced in 1980s. See William Golding, *Lord of
the Flies. Read by the author* (Listening Library, 2005).

2   John Carey, *William Golding. The Man Who Wrote Lord of the Flies* (London, 2010), p. 150.

3   William Golding, *The Hot Gates* (London, 1965), p. 87.

4   Arthur Krystal (ed.), *A Company of Readers. Uncollected Writings of W. H. Auden, Jacques Barzun and Lionel Trilling* (2001), p. 159.

5   Quoted in Carey, *William Golding*, p. 82.

6   Ibid., p. 259.

7   In 'Dit gebeurt er als je gewone kinderen vrijlaat in de wildernis', *De Correspondent* (6 June 2017).

8   Frans de Waal, *The Bonobo and the Atheist*, p. 214.

9   MaryAnn McKibben Dana, 'Friday Link Love: Doubt, Virginia Woolf, and a Real-Life Lord of the Flies', *theblueroom.org* (3 May 2013).

10 Susanna Agnelli, *Street Children. A Growing Urban Tragedy* (London, 1986).

11 Jamie Brown, 'Mates Share 50-Year Bond', *Daily Mercury* (12 December 2014).

12 Quoted in Kay Keavney, 'The Dropout Who Went to Sea', *The Australian Women's Weekly* (19 June 1974).

13 Except where stated otherwise, all quotations by Peter Warner and Mano Totau in this chapter are from my own interviews with them.

14 See in particular Keith Willey, *Naked Island – and Other South Sea Tales* (London, 1970).

15 Steve Bowman, an Australian filmmaker, interviewed David in 2007 and kindly shared his (unpublished) footage with me. This quote is from Bowman's documentary.

16 Willey, *Naked Island*, p. 6.

17 Quoted in Scott Hamilton, 'In remote waters', *readingthemaps. blogspot.com* (18 November 2016).

18 Peter Warner, *Ocean of Light. 30 years in Tonga and the Pacific* (Keerong, 2016), p. 19.

19 This is how Sione remembers it too. 'We stayed very close together,' he told me over the phone. 'Any time there was a quarrel, I tried to calm the boys down. Then they'd cry, they'd apologise, and that was the end of it. That's how it went every time.'

20 Actually, this was lucky. Believing themselves near Samoa, the boys set a course south, whereas in fact they would have had to go north.

21 Willey, *Naked Island*, p. 33.

22 Warner, *Ocean of Light*, p. 89.

23 Charlotte Edwardes, 'Survivor Game Show Based on Public School', *Daily Telegraph* (3 June 2001).

24 Robert Evans and Michael Thot, '5 Ways You Don't Realize Reality Shows Lie', *Cracked.com* (7 July 2014).

25 Girl Scout Research Institute, 'Girls and Reality TV' (2011).

26 Robert Sapolsky, *Behave. The Biology of Humans at Our Best and Worst* (London, 2017), p. 199.

27 Bryan Gibson et al., 'Just "Harmless Entertainment"? Effects of Surveillance Reality TV on Physical Aggression', *Psychology of Popular Media Culture* (18 August 2014).

28 Quoted in CBC Arts, 'George Gerbner Leaves the Mean World Syndrome', *Peace, Earth & Justice News* (8 January 2006).

29 A teacher interviewed by documentary maker Steve Bowman, in the unpublished material he shared with me.

3 The Rise of *Homo puppy*

1 The *Oxford Dictionary* defines hominins as 'A primate of a taxonomic tribe (*Hominini*), which comprises those species regarded as human, directly ancestral to humans, or very closely related to humans.' The broader family of hominids also includes great apes.

2 Charles Darwin, 'To Joseph Dalton Hooker', *Darwin Correspondence Project* (11 January 1844).

3 Richard Dawkins, *The Selfish Gene. 30th Anniversary Edition* (2006), p. ix. Originally published in 1976. Dawkins later removed this passage (see the end of this chapter).

4 Claire Armitstead, 'Dawkins Sees off Darwin in Vote for Most Influential Science Book', *Guardian* (20 July 2017).

5 Michael J. Edwards, 'Fascinating, But at Times I Wish I Could Unread It', review on Amazon.com (7 August 1999). This is one of the highest-rated reviews of the book on Amazon.

6 Marcus E. Raichle and Debra A. Gusnard, 'Appraising the Brain's Energy Budget', *PNAS* (6 August 2002).

7 E. Hermann et al., 'Humans Have Evolved Specialized Skills of Social Cognition: The Cultural Intelligence Hypothesis', *Science* (7 September 2007).

8 Joseph Henrich, *The Secret of Our Success. How Culture Is Driving Human Evolution, Domesticating Our Species, and Making Us Smarter* (Princeton, 2016), pp. 16–17.

9   Ibid., pp. 17–21.

10 Maria Konnikova, *The Confidence Game* (New York, 2016). See the
    epilogue for more about Konnikova's fascinating book.

11 Charles Darwin, *The Expression of the Emotions in Man and Animals*
    (New York, 1872), p. 309. In 2018 a small study with five blue-and-
    yellow macaws was published suggesting this type of parrot also
    has the ability to blush. See Aline Bertin et al., 'Facial Display and
    Blushing: Means of Visual Communication in Blue-and-Yellow
    Macaws (*Ara Ararauna*)?', *PLoS One* (22 August 2019).

12 Johann Carl Fuhlrott, 'Menschliche Überreste aus einer Felsengrotte
    des Düsselthals. Ein Beitrag zur Frage über die Existenz fossiler
    Menschen', in *Verhandlungen des Naturhistorischen Vereins der
    preußischen Rheinlande und Westphalens* (Part 16, 1859), pp. 131–53.

13 The society's name in German was *Niederrheinische Gesellschaft für
    Natur- und Heilkunde*.

14 Paige Madison, 'The Most Brutal of Human Skulls: Measuring and
    Knowing the First Neanderthal', *British Journal for the History of
    Science* (No. 3, 2016), p. 427.

15 This name (*Homo stupidus*) was proposed by biologist Ernst Haeckel,
    but failed to catch on because the anatomist William King had
    already coined *Homo neanderthalensis* two years earlier.

16 Quoted in João Zilhão, 'The Neanderthals: Evolution, Paleoecology,
    and Extinction', in Vicki Cummings, Peter Jordan and Marek
    Zvelebil, *The Oxford Handbook of the Archaeology and Anthropology of
    Hunter-Gatherers* (Oxford, 2014), p. 192.

17 Thomas D. Berger and Erik Trinkaus, 'Patterns of Trauma among
    the Neandertals', *Journal of Archaeological Science* (November 1995).

18 Thomas Wynn and Frederick L. Coolidge, *How to Think Like
    a Neanderthal* (Oxford, 2012), p. 19. If you're still imagining
    Neanderthals as a barbaric kind of caveman, think again. In 2018,
    a team of archaeologists compared the number of skull fractures
    in 295 Neanderthals to *Homo sapiens* (our direct ancestors) from
    the same period. What did they find? No difference. Life for a
    Neanderthal was no more barbaric than it was for us. We, too, seem
    to have been something of the primeval rodeo buck. See Judith Beier
    et al., 'Similar Cranial Trauma Prevalence among Neanderthals and
    Upper Palaeolithic modern humans', *Nature* (14 November 2018).

19 Paola Villa and Wil Roebroeks, 'Neandertal Demise: An
    Archaeological Analysis of the Modern Human Superiority
    Complex', *PLoS One* (30 April 2014).

20 Yuval Noah Harari, *Sapiens. A Brief History of Humankind* (London, 2011), p. 19.

21 Jared Diamond, 'A Brand-New Version of Our Origin Story', *New York Times* (20 April 2018).

22 Except where stated otherwise, my main source for this story is Lee Alan Dugatkin and Lyudmila Trut, *How to Tame a Fox (and Build a Dog). Visionary Scientists and a Siberian Tale of Jump-Started Evolution* (Chicago, 2017).

23 Lee Alan Dugatkin and Lyudmila Trut, 'How to Tame a Fox and Build a Dog', *American Scientist* (No. 4, 2017).

24 Dugatkin and Trut, *How to Tame a Fox*, p. 58.

25 Ibid., p. 124.

26 Robert L. Cieri et al., 'Craniofacial Feminization, Social Tolerance, and the Origins of Behavioral Modernity', *Current Anthropology* (No. 4, 2014).

27 Humans are not the direct descendants of Neanderthals (though given that many people have Neanderthal DNA, *Homo sapiens* and *Homo neanderthalensis* clearly had children together); however, our undomesticated *Homo sapien* ancestors of fifty thousand years ago much more closely resembled Neanderthals, meaning they looked significantly more masculine. See Brian Hare, 'Survival of the Friendliest: *Homo sapiens* Evolved via Selection for Prosociality', *Annual Review of Psychology* (2017).

28 Brian Hare, *The Genius of Dogs. Discovering the Unique Intelligence of Man's Best Friend* (London, 2013). p. 40

29 Ibid., p. 88.

30 Brian Hare, 'Survival of the Friendliest – Brian Hare, Duke Forward in Houston', YouTube (20 January 2016). Hare says this at 3:56 minutes.

31 Domestication has an effect on melanin expression in fur, which also explains the white spots on Dmitri's foxes. Brian Hare, 'Survival of the Friendliest: *Homo sapiens* Evolved via Selection for Prosociality', *Annual Review of Psychology* (2017).

32 Ricardo Miguel Godinho, Penny Spikins and Paul O'Higgins, 'Supraorbital Morphology and Social Dynamics in Human Evolution', *Nature Ecology & Evolution* (No. 2, 2018). Also see Matteo Zanella, 'Dosage analysis of the 7q11.23 Williams region identifies BAZ1B as a major human gene patterning the modern human face and underlying self-domestication', *Science Advances* (4 December 2019).

33 Henrich, *The Secret of Our Success*, p. 214.
34 James Thomas and Simon Kirby, 'Self domestication and the evolution of language', *Biology & Philosophy* (27 March 2018).
35 Peter Turchin, *Ultrasociety. How 10,000 Years of War Made Humans the Greatest Cooperators on Earth* (Chaplin, 2016), p. 48.
36 Joris Luyendijk, 'Parasitair', *NRC Handelsblad* (13 December 2012).
37 Julia Carrie Wong, 'Uber's "hustle-oriented" culture becomes a black mark on employees' résumés', *Guardian* (7 March 2017).
38 Jeremy Lent, *The Patterning Instinct. A Cultural History of Humanity's Search for Meaning* (New York, 2017), pp. 94–5.
39 Julianne Holt-Lunstad, 'Testimony before the US Senate Aging Committee', *aging.senate.gov* (27 April 2017).
40 Helen Louise Brooks, 'The Power of Support from Companion Animals for People Living with Mental Health Problems: A Systematic Review and Narrative Synthesis of the Evidence', *BMC Psychiatry* (5 February 2018).
41 In the late 1980s, evolutionary anthropologist David Buss conducted a survey in thirty-seven countries, asking tens of thousands of people what they look for in a mate. Their responses revealed a slight divergence between the sexes. Looks are more important to men, money matters more to women. Naturally, this was splashed all over the media. What got completely ignored was the one characteristic that was the unanimous number one: kindness. See Dacher Keltner, 'The Compassionate Species', *Greater Good Magazine* (31 July 2012).

4 Colonel Marshall and the Soldiers Who Wouldn't Shoot

1 Quoted in Melyssa Allen, 'Dog Cognition Expert Brian Hare Visits Meredith', *meredith.edu* (October 2016).
2 Carsten K. W. De Dreu et al., 'The Neuropeptide Oxytocin Regulates Parochial Altruism in Intergroup Conflict Among Humans', *Science* (11 June 2010).
3 Raymond Dart, 'The Predatory Transition from Ape to Man', *International Anthropological and Linguistic Review* (No. 4, 1953).
4 Ibid.
5 Quoted in Rami Tzabar, 'Do Chimpanzee Wars Prove That Violence Is Innate?' *bbc.com* (11 August 2015).
6 Richard Wrangham and Dale Peterson, *Demonic Males: Apes and the Origins of Human Violence* (New York, 1996), p. 63.

7   The '!' stands for a clicking sound that is part of the !Kung language.

8   Richard Lee, *The !Kung San* (New York, 1979), p. 398.

9   Steven Pinker, *The Better Angels of Our Nature. Why Violence Has Declined* (London, 2011), p. 36.

10 Ibid., p. xxi.

11 Ibid.

12 On the Battle of Makin, see Anthony King, *The Combat Soldier. Infantry Tactics and Cohesion in the Twentieth and Twenty-First Centuries* (Oxford, 2013), pp. 46–8.

13 Bill Davidson, 'Why Half Our Combat Soldiers Fail to Shoot', *Collier's Weekly* (8 November 1952).

14 Quoted in King, *The Combat Soldier*, p. 48.

15 S. L. A. Marshall, *Men Against Fire. The Problem of Battle Command* (Oklahoma, 2000), p. 79.

16 Ibid., p. 78.

17 Quoted in John Douglas Marshall, *Reconciliation Road: A Family Odyssey* (Washington DC, 2000), p. 190.

18 Ibid.

19 David Lee, *Up Close and Personal: The Reality of Close-Quarter Fighting in World War II* (London, 2006), p. 19.

20 Quoted in Max Hastings, 'Their Wehrmacht Was Better Than Our Army', *Washington Post* (5 May 1985).

21 Richard Holmes, *Acts of War. Behaviour of Men in Battle* (London, 1985), p. 376.

22 Dave Grossman, *On Killing. The Psychological Cost of Learning to Kill in War and Society* (New York, 2009), p. 31.

23 R. A. Gabriel, *No More Heroes. Madness and Psychiatry in War* (New York, 1987), p. 31.

24 Major T. T. S. Laidley, 'Breech-loading Musket', in *The United States Service Magazine* (January 1865), p. 69.

25 Grossman, *On Killing*, pp. 23–6.

26 Ibid., p. 23.

27 George Orwell, *Homage to Catalonia* (London, 2000), p. 39. Originally published in 1938.

28 Randall Collins, *Violence. A Micro-sociological Theory* (Princeton, 2008), p. 53.

29 Ibid., p. 11.

30 Quoted in Craig McGregor, 'Nice Boy from the Bronx?', *New York Times* (30 January 1972).

31 Lee Berger, 'Brief Communication: Predatory Bird Damage to the Taung Type-Skull of Australopithecus africanus Dart 1925', *American Journal of Physical Anthropology* (31 May 2006).

32 On this debate, see John Horgan, 'Anthropologist Brian Ferguson Challenges Claim that Chimp Violence is Adaptive', *Scientific American* (18 September 2014).

33 Michael L. Wilson et al., 'Lethal Aggression in Pan is Better Explained by Adaptive Strategies than Human Impacts', *Nature* (18 September 2014).

34 Brian Hare, 'Survival of the Friendliest: *Homo sapiens* Evolved via Selection for Prosociality', *Annual Review of Psychology* (2017), pp. 162–3.

35 Robert Sapolsky, 'Rousseau with a Tail. Maintaining a Tradition of Peace Among Baboons', in *War, Peace, and Human Nature. The Convergence of Evolutionary and Cultural Views* (Oxford, 2013), p. 421.

36 John Horgan, 'The Weird Irony at the Heart of the Napoleon Chagnon Affair', *Scientific American* (18 February 2013).

37 Robert Sapolsky, *Behave. The Biology of Humans at Our Best and Worst* (London, 2017), p. 314.

38 R. Brian Ferguson, 'Born to Live: Challenging Killer Myths', in Robert W. Sussman and C. Robert Cloninger (eds), *Origins of Altruism and Cooperation* (New York, 2009), pp. 258–9.

39 Quoted in Christopher Ryan and Cacilda Jethá, *Sex at Dawn. How We Mate, Why We Stray, and What It Means for Modern Relationships* (New York, 2010), p. 196.

40 Douglas Fry, 'War, Peace, and Human Nature: The Challenge of Achieving Scientific Objectivity', in Douglas Fry (ed.), *War, Peace, and Human Nature. The Convergence of Evolutionary and Cultural Views* (Oxford, 2013), pp. 18–19.

41 Ibid., p. 20.

42 Douglas P. Fry and Patrik Söderberg, 'Lethal Aggression in Mobile Forager Bands and Implications for the Origins of War', *Science* (19 July 2013).

43 Kim R. Hill et al., 'Hunter-Gatherer Inter-Band Interaction Rates. Implications for Cumulative Culture', *PLoS One* (24 June 2014).

44 K. R. Hill et al., 'Co-residence Patterns in Hunter-Gatherer Societies Show Unique Human Social Structure', *Science* (11 March 2011). See also Coren L. Apicella, 'Social networks and cooperation in hunter-gatherers', *Nature* (26 January 2012).

45 Jonathan Haas and Matthew Piscitelli, 'The Prehistory of Warfare. Misled by Ethnography', in Douglas Fry (ed.), *War, Peace, and Human Nature*, pp. 178–81.

46 Ibid., pp. 181–3.

47 Two digs are consistently cited as offering the first 'proof' of prehistoric warfare. The first is Jebel Sahaba in northern Sudan, where in 1964 archaeologists found sixty-one skeletons dating back approximately thirteen thousand years, of which twenty-one bore traces of a violent death. More recent analysis reduces this number to four. See Robert Jurmain, 'Paleoepidemiolgical Patterns of Trauma in a Prehistoric Population from Central California', *American Journal of Physical Anthropology* (12 April 2001). The people of Jebel Sahaba lived on the fertile banks of the Nile and built a necropolis for their dead, making it probable that they were already living in settlements. The second frequently mentioned site is Naturuk, near Lake Turkana in Kenya, where twenty-seven skeletons (bearing traces of violence) were found, estimated to be ten thousand years old. When archaeologists published this discovery in *Nature* in 2016, global media seized on it as the definitive 'proof' that humans are innately warlike creatures. But the significance of the Naturuk find is still contested. Numerous archaeologists have pointed out that the banks of Lake Turkana were a fertile location where hunter-gatherers converged, making it quite plausible that they had already consolidated their belongings and given up their nomadic lifestyle. A few months after this article appeared, *Nature* published a reaction from another team of archaeologists who questioned the validity of the conclusion that these 'victims' had met a violent end at all. This latter article went ignored by most media. See Christopher M. Stojanowski et al., 'Contesting the Massacre at Nataruk', *Nature* (24 November 2016). Even without these controversies, it is important to realise that – apart from Jebel Sahaba and Naturuk – there is no evidence for any prehistoric war, in stark contrast to the surfeit of uncontroversial archaeological evidence (in the form of cave paintings and mass graves) for war in the period *after* humans began living in permanent settlements and farming.

48 R. Brian Ferguson, 'Pinker's List. Exaggerating Prehistoric War Mortality', in Douglas Fry (ed.), *War, Peace, and Human Nature*, pp. 126. See also Hisashi Nakao et al., 'Violence in the Prehistoric Period of Japan: The Spatio-Temporal Pattern of Skeletal Evidence for Violence in the Jomon Period', *Biology Letters* (1 March 2016).

5 The Curse of Civilisation

1 Quoted in Sarah Blaffer Hrdy, *Mothers and Others. The Evolutionary Origins of Mutual Understanding* (2009), p. 27.
2 Catherine A. Lutz, *Unnatural Emotions: Everyday Sentiments on a Micronesian Atoll & Their Challenge to Western Theory* (Chicago, 1988).
3 Christopher Boehm, *Hierarchy in the Forest. The Evolution of Egalitarian Behavior* (Cambridge, 1999), p. 68. Also see Christopher Boehm, *Moral Origins. The Evolution of Virtue, Altruism and Shame* (New York, 2012), pp. 78–82.
4 The '!' stands for a clicking sound that is part of the !Kung language.
5 Robert Lee, *The !Kung San: Men, Women, and Work in a Foraging Society* (Cambridge, 1979), p. 244.
6 Ibid., p. 246.
7 Quoted in Blaffer Hrdy, *Mothers and Others*, p. 27.
8 Lee, *The !Kung San*, pp. 394–5.
9 It probably wouldn't have been possible to rein in overconfident chieftains without the simple yet effective innovation of projectile weapons, which is to say we learned to throw stones, hurl spears and shoot arrows. Comparison of excavated *Homo* skeletons shows that, over time, our shoulders and wrists evolved, making us better pitchers. Whereas humans have very good aim, chimpanzees and orangutans do not (an angry chimp might occasionally throw things, but usually misses). Archaeologists think our projectile weapons were probably quite a bit more refined than anything Neanderthals had. According to evolutionary anthropologist Peter Turchin, this should be considered the pivotal invention in the history of humankind, even outranking fire, farming and the wheel. Without projectile weapons, more aggressive members of our species would have had significantly more offspring and *Homo puppy* could never have domesticated itself.
10 Individually, foragers prefer the company of their own kin. If men have sole authority, they favour their own families. If men and women share authority, however, they have to compromise. They'll want to live with both sides of the family, resulting in a more complex social network. This is exactly what we see among nomadic hunter-gatherers. See M. Dyble et al., 'Sex Equality Can Explain the Unique Social Structure of Hunter-Gatherer Bands', *Science*, Vol. 348, Issue 6236 (15 May 2015). Also see Hannah Devlin, 'Early Men and Women Were Equal, Say Scientists', *Guardian* (14 May 2015).

11 Blaffer Hrdy, *Mothers and Others*, p. 128.

12 Ibid., p. 134.

13 Nicholas A. Christakis, *Blueprint. The Evolutionary Origins of a Good Society* (New York, 2019), pp. 141–3.

14 Carel van Schaik and Kai Michel, *The Good Book of Human Nature. An Evolutionary Reading of the Bible* (New York, 2016), p. 51.

15 Which is not to say that 1960s-era hippies were right in thinking humans were made for free love. Marriage dovetails perfectly with our nature, *Homo puppy* being one of the few mammals to engage in 'pair binding' – aka romantic love. Granted, we're not all heroes at staying faithful till death do us part, but the science says that a loving relationship is a universal human desire. See Christakis, *Blueprint*, p. 168.

16 Quoted in E. Leacock, *Myths of Male Dominance. Collected Articles on Women Cross-Culturally* (New York, 1981), p. 50.

17 Jared Diamond, *The World Until Yesterday. What Can We Learn from Traditional Societies?* (London, 2013), p. 11.

18 In north-eastern Louisiana is an archaeological site with finds dating back 3,200 years in terrain filled with mounds that can only be manmade. The largest, Bird Mound, is seventy-two feet high and would require eight million baskets of sand weighing fifty-five pounds each to create. Archaeological research has shown that its construction took no more than a few months and was the concerted labour of at least ten thousand workers. See Anthony L. Ortmann, 'Building Mound A at Poverty Point, Louisiana: Monumental Public Architecture, Ritual Practice, and Implications for Hunter-Gatherer Complexity', *Geoarcheology* (7 December 2012).

19 Jens Notroff, Oliver Dietrich and Klaus Schmidt, 'Building Monuments, Creating Communities. Early Monumental Architecture at Pre-Pottery Neolithic Göbekli Tepe' in James F. Osborne (ed.), *Approaching Monumentality in Archeology* (New York, 2014), pp. 83–105.

20 Erik Trinkaus et al., *The People of Sunghir: Burials, Bodies, and Behavior in the Earlier Upper Paleolithic* (Oxford, 2014).

21 David Graeber and David Wengrow, 'How to Change the Course of Human History (at Least, the Part That's Already Happened)', *Eurozine* (2 March 2018).

22 The term 'snuggle for surival' was coined by biologist Martin Nowak. See Martin Nowak, 'Why We Help', *Scientific American* (No. 1, 2012), pp. 34–9.

23   Van Schaik and Michel, *The Good Book of Human Nature*, pp. 44–5.
24   Ibid., pp. 48–9.
25   Gregory K. Dow, Leanna Mitchell and Clyde G. Reed, 'The Economics of Early Warfare over Land', *Journal of Development Economics* (July 2017). The second section of this article contains a good overview of the archaeological evidence.
26   Douglas W. Bird et al., 'Variability in the Organization and Size of Hunter-Gatherer Groups. Foragers Do Not Live in Small-Scale Societies', *Journal of Human Evolution* (June 2019).
27   Turchin, *Ultrasociety*, p. 163.
28   R. Brian Ferguson, 'Born to Live: Challenging Killer Myths', in Robert W. Sussman and C. Robert Cloninger (eds), *Origins of Altruism and Cooperation* (New York, 2009), pp. 265–6.
29   Genesis 3:19–24. See also Van Schaik and Michel, *The Good Book of Human Nature*, pp. 44–5.
30   Ibid., pp. 50–51.
31   Jared Diamond authored the classic article about how we screwed up by inventing agriculture. See Jared Diamond, 'The Worst Mistake in the History of the Human Race', *Discover Magazine* (May 1987).
32   James C. Scott, *Against the Grain. A Deep History of the Earliest States* (New Haven, 2017), pp. 104–5.
33   Jean-Jacques Rousseau, *A Dissertation On the Origin and Foundation of The Inequality of Mankind and is it Authorised by Natural Law?* Originally published in 1754.
34   Van Schaik and Michel, *The Good Book of Human Nature*, pp. 52–4.
35   Hervey C. Peoples, Pavel Duda and Frank W. Marlowe, 'Hunter-Gatherers and the Origins of Religion', *Human Nature* (September 2016).
36   Frank Marlowe, *The Hadza. Hunter-Gatherers of Tanzania* (Berkeley, 2010), p. 61.
37   Ibid., pp. 90–93.
38   Quoted in Lizzie Wade, 'Feeding the gods: Hundreds of skulls reveal massive scale of human sacrifice in Aztec capital', *Science* (21 June 2018).
39   Quoted in Richard Lee, 'What Hunters Do for a Living, or, How to Make Out on Scarce Resources', *Man the Hunter* (Chicago, 1968), p. 33.
40   James C. Scott, *Against the Grain*, pp. 66–7.
41   Turchin, *Ultrasociety*, pp. 174–5.
42   Scott, *Against the Grain*, pp. 27–9.

43 For an extensive historical overview, see David Graeber, *Debt. The First 5,000 Years* (London, 2011).

44 Scott, *Against the Grain*, pp. 139–49.

45 Ibid., p. 162.

46 Owen Lattimore, 'The Frontier in History', in *Studies in Frontier History: Collected Papers, 1928–1958* (London, 1962), pp. 469–91.

47 Quoted in Bruce E. Johansen, *Forgotten Founders* (Ipswich, 1982), Chapter 5.

48 James W. Loeven, *Lies My Teacher Told Me. Everything Your American History Textbook Got Wrong* (2005), pp. 101–2.

49 Quoted in Junger, *Tribe*, pp. 10–11.

50 Ibid., pp. 14–15.

51 Iconic in this genre is Edward Gibbon's book *The Decline and Fall of the Roman Empire* (1776). A modern-day bestseller is *Collapse* (2005) by Jared Diamond.

52 Some scholars question whether the *Iliad* and the *Odyssey* should be attributed to an individual at all, suggesting the name Homer should be seen as a label attached to a good Greek tale. This would mean Homer never existed as such.

53 Adam Hochschild, *Bury the Chains: Prophets and Rebels in the Fight to Free an Empire's Slaves* (Boston, 2005), p. 2.

54 Max Roser and Esteban Ortiz-Ospina, 'Global Extreme Poverty', *OurWorldInData.org* (2018).

55 This is the opening sentence of Rousseau's book *The Social Contract* (originally published in 1762).

56 Bjørn Lomborg, 'Setting the Right Global Goals', *Project Syndicate* (20 May 2014).

57 Max Roser and Esteban Ortiz-Ospina, 'Global Extreme Poverty'.

58 Quoted in Chouki El Hamel, *Black Morocco. A History of Slavery, Race, and Islam* (Cambridge, 2013), p. 243.

59 Mauritius in West Africa was the last country in the world to abolish slavery, in 1981.

60 In Persian and Roman times, state expansion was already making the world incrementally safer. Though it sounds like a paradox, there is a logical explanation. As countries and empires grew, more of their citizens lived further away from borders. It was at borders that wars were fought; life was more peaceful in the interior. Illustrative of this is the *Pax Romana* (Roman Peace), a long period of stability secured by the great campaigns of the mightiest Leviathans. In this sense, at least, Hobbes was right: better one all-powerful emperor

than a hundred frustrated petty kings. See Turchin, *Ultrasociety*, pp. 201–2.

61   José María Gómez et al., 'The Phylogenetic Roots of Human Lethal Violence, Supplementary Information', *Nature* (13 October 2016), p. 9.

62   In 2017, 2,813,503 deaths were registered in the United States. According to the National Violent Death Reporting System, 19,500 of these were victims of homicide. That same year, 150,214 deaths were registered in the Netherlands, of which 158 were homicides.

63   This story is probably apocryphal. 'Not letting the facts ruin a good story', *South China Morning Post* (29 September 2019).

6 The Mystery of Easter Island

1   My account of Roggeveen's life and expedition is based on the excellent biography by Roelof van Gelder, *Naar het aards paradijs. Het rusteloʒe leven van Jacob Roggeveen, ontdekker van Paaseiland (1659–1729)* (Amsterdam, 2012).

2   F. E. Baron Mulert, *De reis van Mr. Jacob Roggeveen ter ontdekking van het Zuidland (1721–1722)*, (The Hague, 1911), p. 121.

3   H. J. M. Claessen, 'Roggeveen zag geen reuzen toen hij Paaseiland bezocht', *NRC Handelsblad* (18 April 2009).

4   This Swiss hotel manager was Erich von Däniken, and his book was titled *Chariots of the Gods? Unsolved Mysteries of the Past*.

5   Lars Fehren-Schmitz, 'Genetic Ancestry of Rapanui before and after European Contact', *Current Biology* (23 October 2017).

6   Katherine Routledge, *The Mystery of Easter Island. The Story of an Expedition* (London, 1919).

7   Reidar Solsvik, 'Thor Heyerdahl as world heritage', *Rapa Nui Journal* (May 2012).

8   Quoted in Jo Anne Van Tilburg, 'Thor Heyerdahl', *Guardian* (19 April 2002).

9   William Mulloy, 'Contemplate The Navel of the World', *Rapa Nui Journal* (No. 2, 1991). Originally published in 1974.

10   Jared Diamond, *Collapse. How Societies Choose to Fail or Succeed* (New York, 2005), p. 109.

11   J. R. Flenley and Sarah M. King, 'Late Quaternary Pollen Records from Easter Island', *Science* (5 January 1984).

12 Diamond owes a debt to historian Clive Ponting, who wrote about Easter Island in his book *A Green History of the World* (1991). On the first page, Ponting describes the island as Roggeveen found it: 'about 3,000 people living in squalid reed huts or caves, engaged in almost perpetual warfare and resorting to cannibalism in a desperate attempt to supplement the meagre food supplies available on the island'.

13 Paul Bahn and John Flenley, *Easter Island, Earth Island* (London, 1992).

14 Jan J. Boersema, *The Survival of Easter Island. Dwindling Resources and Cultural Resilience* (Cambridge, 2015).

15 Carlyle Smith, 'The Poike Ditch', in Thor Heyerdahl (ed.), *Archeology of Easter Island. Reports of the Norwegian Archaeological Expedition to Easter Island and the East Pacific* (Part 1, 1961), pp. 385–91.

16 Carl P. Lipo and Terry L. Hunt, 'A.D. 1680 and Rapa Nui Prehistory', *Asian Perspectives* (No. 2, 2010). Also see Mara A. Mulrooney et al., 'The myth of A.D. 1680. New Evidence from Hanga Ho'onu, Rapa Nui (Easter Island)', *Rapa Nui Journal* (October 2009).

17 Caroline Polet, 'Indicateurs de stress dans un échantillon d'anciens Pascuans', *Antropo* (2006), pp. 261–70.

18 See Vincent H. Stefan et al. (ed.), *Skeletal Biology of the Ancient Rapanui (Easter Islanders)*, (Cambridge, 2016).

19 Carl P. Lipo et al., 'Weapons of War? Rapa Nui Mata'a Morphometric Analyses', *Antiquity* (February 2016), pp. 172–87.

20 Quoted in Kristin Romey, 'Easter Islanders' Weapons Were Deliberately Not Lethal', *National Geographic* (22 February 2016).

21 Terry L. Hunt and Carl P. Lipo, 'Late Colonization of Easter Island', *Science* (17 March 2006).

22 Ronald Wright, *A Short History of Progress* (Toronto, 2004), p. 61.

23 Hans-Rudolf Bork and Andreas Mieth, 'The Key Role of the *Jubaea* Palm Trees in the History of Rapa Nui: a Provocative Interpretation', *Rapa Nui Journal* (October 2003).

24 Nicolas Cauwe, 'Megaliths of Easter Island', *Proceedings of the International Conference 'Around the Petit-Chausseur Sit'* (Sion, 2011).

25 The archaeologists Carl Lipo and Terry Hunt think some statues were 'walked' vertically into place using ropes, not trees, the same way you might move a refrigerator or washing machine. This method also requires fewer people. See Carl Lipo and Terry Hunt, *The Statues that Walked. Unraveling the Mystery of Easter Island*

(New York, 2011). Lipo and Hunt's story was popular in the media, but Jan Boersema still believes most of the statues were rolled on tree trunks by big groups of people, because efficiency was not the motivating factor behind such collective work events.

26 E. E. W. Schroeder, *Nias. Ethnographische, geographische en historische aanteekeningen en studien* (Leiden, 1917).

27 S. S. Barnes, Elizabeth Matisoo-Smith and Terry L. Hunt, 'Ancient DNA of the Pacific Rat (*Rattus exulans*) from Rapa Nui (Easter Island)', *Journal of Archaeological Science* (Vol. 33, November 2006).

28 Mara A. Mulrooney, 'An Island-Wide Assessment of the Chronology of Settlement and Land Use on Rapa Nui (Easter Island) Based on Radiocarbon Data', *Journal of Archaeological Science* (No. 12, 2013). Didn't the rats pose a problem for farming on the island? Boersema thinks not. 'Most food crops were tubers,' he explains, 'which grow beneath the soil. And the bananas grew on small trees which made them less appealing to rats.'

29 Quoted in 'Easter Island Collapse Disputed By Hawaii Anthropologist', *Huffington Post* (6 December 2017).

30 Jacob Roggeveen, *Dagverhaal der ontdekkings-reis van Mr. Jacob Roggeveen* (Middelburg, 1838), p. 104.

31 Bolton Glanvill Corney, *The Voyage of Captain Don Felipe González to Easter Island 1770–1* (Cambridge, 1908), p. 93.

32 Beverley Haun, *Inventing Easter Island* (Toronto, 2008), p. 247.

33 James Cook, *A Voyage Towards the South Pole and Round the World*, Part 1 (1777).

34 Henry Lee, 'Treeless at Easter', *Nature* (23 September 2004).

35 The book in question is Thor Heyerdahl et al., *Archaeology of Easter Island. Reports of the Norwegian Archaeological Expedition to Easter Island and the East Pacific* (Part 1, 1961), p. 51.

36 Thor Heyerdahl, *Aku-Aku: The Secret of Easter Island* (1957).

37 Carl Behren's account is included as an appendix to Glanvill Corney, *The voyage of Captain Don Felipe González to Easter Island 1770–1*, p. 134.

38 Cook, *A Voyage Towards the South Pole and Round the World*, Chapter 8.

39 Some scientists believe the statues fell down during an earthquake. Others think some *moai* were laid down over the graves of deceased chieftains. See Edmundo Edwards et al., 'When the Earth Trembled, the Statues Fell', *Rapa Nui Journal* (March 1996).

40 This also gave rise to the 'Birdman Cult', an annual competition between young men representing different tribes to snatch the first

sooty tern (a seabird) egg of the season. Exactly when this tradition arose is unknown, but it was probably before Roggeveen's arrival. This cult was also linked to the *moai*. After the competition, the newly elected leader went to live in a house outside the stone quarry where the statues were carved. When Roggeveen arrived in 1722, the *moai* still had a definite ceremonial function, even if it was no longer possible to transport them (using trees) and even though the Birdman Cult probably already existed.

41 Josh Pollard, Alistair Paterson and Kate Welham, 'Te Miro o'one: the Archaeology of Contact on Rapa Nui (Easter Island)', *World Archaeology* (December 2010).

42 Henry Evans Maude, *Slavers in Paradise: The Peruvian Labour Trade in Polynesia, 1862–1864* (Canberra, 1981), p. 13.

43 Nicolas Casey, 'Easter Island Is Eroding', *New York Times* (20 July 2018).

7 In the Basement of Stanford University

1 Quoted in Ben Blum, 'The Lifespan of a Lie', *Medium.com* (7 June 2018).

2 Craig Haney, Curtis Banks and Philip Zimbardo, 'A Study of Prisoners and Guards in a Simulated Prison', *Naval Research Review* (1973).

3 Malcolm Gladwell, *The Tipping Point. How Little Things Can Make A Big Difference* (London, 2000), p. 155.

4 Haney, Banks and Zimbardo, 'A Study of Prisoners and Guards in a Simulated Prison'.

5 Muzafer Sherif, *Group Conflict and Co-operation. Their Social Psychology* (London, 2017), p. 85. Originally published in 1967.

6 Muzafer Sherif et al., *The Robbers Cave Experiment. Intergroup Conflict and Cooperation* (Middletown, 1988), p. 115.

7 Ibid., p. 98.

8 Quoted in Gina Perry, *The Lost Boys. Inside Muzafer Sherif's Robbers Cave Experiment* (London, 2018), p. 39.

9 Ibid. p. 138.

10 Ibid. p. 139.

11 Ibid. p. 146.

12 In the Stanford Prison Experiment twelve students were assigned the role of prisoner (nine plus three stand-ins), and twelve that of guard (nine plus three stand-ins).

13   Quoted in Blum, 'The Lifespan of a Lie'.

14   Philip Zimbardo, *The Lucifer Effect. How Good People Turn Evil* (London, 2007), p. 55.

15   Peter Gray, 'Why Zimbardo's Prison Experiment Isn't in My Textbook', *Psychology Today* (19 October 2013).

16   Quoted in Romesh Ratnesar, 'The Menace Within', *Stanford Magazine* (July/August 2011).

17   Dave Jaffe, 'Self-perception', *Stanford Prison Archives*, No. ST-b09-f40.

18   'Tape 2' (14 August 1971), *Stanford Prison Archives*, No. ST-b02-f02.

19   A. Cerovina, 'Final Prison Study Evaluation' (20 August 1971), No. ST-b09-f15.

20   'Tape E' (no date), No. ST-b02-f21, pp. 1–2.

21   Quoted in Blum, 'The Lifespan of a Lie'.

22   Blum, 'The Lifespan of a Lie'.

23   Ibid.

24   Ibid.

25   Quoted in Alastair Leithead, 'Stanford prison experiment continues to shock', *BBC* (17 August 2011).

26   For many years, psychologists used Zimbardo's 'experiment' to spark students' enthusiasm for the field. Thibault Le Texier spoke to a number of lecturers who said they liked discussing the Stanford Prison Experiment because it at least got students to look up from their phones. In response to my question whether it should still be taught in classrooms today, Le Texier answered dryly, 'The Stanford Experiment is a pretty good overview of all the errors you can make in scientific research.'

27   Quoted in: Kim Duke and Nick Mirsky, 'The Stanford Prison Experiment,' *BBC Two* (11 May 2002). Dave Eshelman's full quote in this documentary is: 'It would have been interesting to see what would have happened had I not decided to force things. [...] We'll never know.'

28   Emma Brockes, 'The Experiment', *Guardian* (16 October 2001).

29   Ibid.

30   Graeme Virtue, 'Secret service; What happens when you put good men in an evil place and film it for telly? Erm, not that much actually', *Sunday Herald* (12 May 2002).

31   Blum, 'The Lifespan of a Lie'.

8 Stanley Milgram and the Shock Machine

1   'Persons Needed for a Study of Memory" *New Haven Register* (18 June 1961).
2   Stanley Milgram, *Obedience to Authority. An Experimental View* (London, 2009), pp. 30–31. Originally published in 1974.
3   Stanley Milgram, 'Behavioral Study of Obedience', *Journal of Abnormal and Social Psychology*, Vol. 67, Issue 4 (1963).
4   Walter Sullivan, 'Sixty-five Percent in Test Blindly Obey Order to Inflict Pain', *New York Times* (26 October 1963).
5   Milgram, *Obedience to Authority*, p. 188.
6   Milgram said this in an interview on the television programme *Sixty Minutes* on 31 March 1979.
7   Quoted in Amos Elon, 'Introduction', in Hannah Arendt, *Eichmann in Jerusalem. A Report on the Banality of Evil* (London, 2006), p. xv. Originally published in 1963.
8   Arendt, *Eichmann in Jerusalem*.
9   Quoted in Harold Takooshian, 'How Stanley Milgram Taught about Obedience and Social Influence', in Thomas Blass (ed.), *Obedience to Authority* (London, 2000), p. 10.
10 Quoted in Gina Perry, *Behind the Shock Machine. The Untold Story of the Notorious Milgram Psychology Experiments* (New York, 2013), p. 5.
11 Ibid., p. 327.
12 Ibid., p. 134.
13 Gina Perry, 'The Shocking Truth of the Notorious Milgram Obedience Experiments', *Discover Magazine* (2 October 2013).
14 Milgram, 'Behavioral Study of Obedience'.
15 Perry, *Behind the Shock Machine* (2012), p. 164. See also Gina Perry et al., 'Credibility and Incredulity in Milgram's Obedience Experiments: A Reanalysis of an Unpublished Test', *Social Psychology Quarterly* (22 August 2019).
16 Stanley Milgram, 'Evaluation of Obedience Research: Science or Art?' *Stanley Milgram Papers* (Box 46, file 16). Unpublished manuscript (1962).
17 Quoted in Stephen D. Reicher, S. Alexander Haslam and Arthur Miller, 'What Makes a Person a Perpetrator? The Intellectual, Moral, and Methodological Arguments for Revisiting Milgram's Research on the Influence of Authority', *Journal of Social Issues*, Vol. 70, Issue 3 (2014).

18   Quoted in Perry, *Behind the Shock Machine*, p. 93.

19   Quoted in Cari Romm, 'Rethinking One of Psychology's Most Infamous Experiments', *The Atlantic* (28 January 2015).

20   Stephen Gibson, 'Milgram's Obedience Experiments: a Rhetorical Analysis', *British Journal of Social Psychology*, Vol. 52, Issue 2 (2011).

21   S. Alexander Haslam, Stephen D. Reicher and Megan E. Birney, 'Nothing by Mere Authority: Evidence that in an Experimental Analogue of the Milgram Paradigm Participants are Motivated not by Orders but by Appeals to Science', *Journal of Social Issues*, Vol. 70, Issue 3 (2014).

22   Quoted in Perry, *Behind the Shock Machine*, p. 176.

23   Quoted in S. Alexander Haslam and Stephen D. Reicher, 'Contesting the "Nature" of Conformity: What Milgram and Zimbardo's Studies Really Show', *PLoS Biology*, Vol. 10, Issue 11 (2012).

24   Quoted in Perry, *Behind the Shock Machine*, p. 70.

25   Quoted in Blum, 'The Lifespan of a Lie'.

26   Ibid.

27   Quoted in 'Tape E' (no date), *Stanford Prison Archives*, No.: ST-b02-f21, p. 6.

28   Ibid., p. 2.

29   Perry, *Behind the Shock Machine*, p. 240.

30   Arendt, *Eichmann in Jerusalem*, p. 276.

31   Quoted in Bettina Stangneth, *Eichmann Before Jerusalem: The Unexamined Life of a Mass Murderer* (London, 2015).

32   Quoted in 'The Adolph Eichmann Trial 1961', in *Great World Trials* (Detroit, 1997), pp. 332–7.

33   Ian Kershaw, ' "Working Towards the Führer." Reflections on the Nature of the Hitler Dictatorship', *Contemporary European History*, Vol. 2, Issue 2 (1993).

34   See, for example, Christopher R. Browning, 'How Ordinary Germans Did It', *New York Review of Books* (20 June 2013).

35   Quoted in Roger Berkowitz, 'Misreading 'Eichmann in Jerusalem', *New York Times* (7 July 2013).

36   Ibid.

37   Ada Ushpiz, 'The Grossly Misunderstood "Banality of Evil" Theory', *Haaretz* (12 October 2016).

38   Quoted in Perry. *Behind the Shock Machine*, p. 72.

39   Matthew M. Hollander, 'The Repertoire of Resistance: Non-Compliance With Directives in Milgram's "Obedience" experiments', *British Journal of Social Psychology*, Vol. 54, Issue 3 (2015).

40　Matthew Hollander, 'How to Be a Hero: Insight From the Milgram Experiment', *Huffington Post* (27 February 2015).

41　Quoted in Bo Lidegaard, *Countrymen: The Untold Story of How Denmark's Jews Escaped the Nazis, of the Courage of Their Fellow Danes – and of the Extraordinary Role of the SS* (New York, 2013), p. 71.

42　Ibid., p. 353.

43　Ibid., p. 113.

44　Ibid., p. 262.

45　Ibid., p. 173.

46　Ibid., p. 58.

47　Peter Longerich, 'Policy of Destruction. Nazi Anti-Jewish Policy and the Genesis of the "Final Solution"', United States Holocaust Memorial Museum, Joseph and Rebecca Meyerhoff Annual Lecture (22 April 1999), p. 5.

48　Lidegaard, *Countrymen*, p. 198.

49　Ibid., p. 353.

9　The Death of Catherine Susan Genovese

1　For this first report of the murder, see Martin Gansberg, '37 Who Saw Murder Didn't Call the Police', *New York Times* (27 March 1964).

2　Nicholas Lemann, 'A Call for Help', *The New Yorker* (10 March 2014).

3　Gansberg, '37 Who Saw Murder Didn't Call the Police', *New York Times*.

4　Peter C. Baker, 'Missing the Story', *The Nation* (8 April 2014).

5　Kevin Cook, *Kitty Genovese. The Murder, The Bystanders, The Crime That Changed America* (New York, 2014), p. 100.

6　Abe Rosenthal, 'Study of the Sickness Called Apathy', *New York Times* (3 May 1964).

7　Gladwell, *The Tipping Point*, p. 27.

8　Rosenthal said this in the documentary *The Witness* (2015), made by Kitty's brother Bill Genovese.

9　Bill Keller, 'The Sunshine Warrior', *New York Times* (22 September 2002).

10　John M. Darley and Bibb Latené, 'Bystander Intervention in Emergencies', *Journal of Personality and Social Psychology*, Vol. 8, Issue 4 (1968).

11   Malcolm Gladwell says 85 and 31 per cent in his book, but the original article makes clear that these are the percentages of people who ran to help before the 'victim's' first call for assistance had ended (after seventy-five seconds). Many people responded after this but still within two and a half minutes.

12   Maureen Dowd, '20 Years After the Murder of Kitty Genovese, the Question Remains: Why?', *New York Times* (12 March 1984).

13   Cook, *Kitty Genovese*, p. 161.

14   Rachel Manning, Mark Levine and Alan Collins, 'The Kitty Genovese Murder and the Social Psychology of Helping. The Parable of the 38 Witnesses', *American Psychologist*, Vol. 62, Issue 6 (2007).

15   Sanne is a pseudonym. Her real name is not known to me, but it is to her four rescuers.

16   'Mannen die moeder en kind uit water redden: "Elke fitte A'dammer zou dit doen"', *at5.nl* (10 February 2016).

17   'Vier helden redden moeder en kind uit zinkende auto', *nos.nl* (10 February 2016).

18   Peter Fischer et al., 'The bystander-effect: a meta-analytic review on bystander intervention in dangerous and non-dangerous emergencies', *Psychological Bulletin*, Vol. 137, Issue 4 (2011).

19   Ibid.

20   R. Philpot et al., 'Would I be helped? Cross-National CCTV Shows that Intervention is the Norm in Public Conflicts', *American Psychologist* (March 2019).

21   This account is based on three books: Kevin Cook, *Kitty Genovese* (2014); Catherine Pelonero, *Kitty Genovese. A True Account of a Public Murder and Its Private Consequences* (New York, 2014); and Marcia M. Gallo, *'No One Helped.' Kitty Genovese, New York City and the Myth of Urban Apathy* (Ithaca, 2015).

22   She said this in Bill Genovese's 2015 documentary *The Witness*.

23   Baker, 'Missing the Story'.

24   Robert C. Doty, 'Growth of Overt Homosexuality In City Provokes Wide Concern', *New York Times* (17 December 1963).

25   Quoted in Pelonero, *Kitty Genovese*, p. 18.

26   Ibid.

27   Ibid.

28   Ibid.

29   Saul M. Kassin, 'The Killing of Kitty Genovese: What Else Does This Case Tell Us?' *Perspectives on Psychological Science*, Vol. 12, Issue 3 (2017).

Part 3 Why Good People Turn Bad

1   For a lucid discussion, see Jesse Bering, 'The Fattest Ape: An
    Evolutionary Tale of Human Obesity', *Scientific American* (2
    November 2010).

10 How Empathy Blinds

1   James Burk, 'Introduction', in James Burk (ed.), *Morris Janowitz. On
    Social Organization and Social Control* (Chicago, 1991).
2   See, for example, Martin Van Creveld, *Fighting Power: German and
    US Army Performance, 1939–1945*, ABC-CLIO (1982).
3   Max Hastings, 'Their Wehrmacht Was Better Than Our Army',
    Washington Post (5 May 1985).
4   Quoted in Edward A. Shils and Morris Janowitz, 'Cohesion and
    Disintegration in the Wehrmacht in World War II', *Public Opinion
    Quarterly*, Vol. 12, Issue 2 (1948).
5   Ibid., p. 281.
6   Ibid., p. 303.
7   Ibid., p. 284.
8   Felix Römer, *Comrades. The Wehrmacht from Within* (Oxford, 2019).
9   Janowitz and Shils' first article would become one of the most widely
    cited studies in post-war sociology. There is broad consensus among
    sociologists for the validity of their 'primary group theory', i.e. the
    notion that soldiers fight primarily for their immediate comrades,
    though with a few caveats. Some scientists point out that there was
    real hatred towards the enemy among ordinary recruits as well,
    particularly on the Eastern Front. Also that, where twenty-first-
    century professional soldiers are concerned, only three factors really
    determine success: training, training and more training. Sociologists
    these days accordingly distinguish between *group cohesion* and *task
    cohesion*, meaning effective collaboration does not require soldiers to
    feel deep affection for each other. Nevertheless, ties of brotherhood
    among enlistees have historically been crucial in the vast majority of
    wars.
10 Quoted in Michael Bond, *The Power of Others. Peer Pressure, Group
    Think, and How the People Around Us Shape Everything We Do*
    (London, 2015), pp. 128–9.

11 Amy Chua, *Political Tribes. Group Instinct and the Fate of Nations* (New York, 2018), p. 100.

12 Bond, *The Power of Others*, pp. 94–5.

13 Quoted in ibid., pp. 88–9.

14 Benjamin Wallace-Wells, 'Terrorists in the Family', *New Yorker* (24 March 2016).

15 Quoted in Donato Paolo Mancini and Jon Sindreu, 'Sibling Ties Among Suspected Barcelona Plotters Underline Trend', *Wall Street Journal* (25 August 2017).

16 Deborah Schurman-Kauflin, 'Profiling Terrorist Leaders. Common Characteristics of Terror Leaders', *Psychology Today* (31 October 2013).

17 Aya Batrawy, Paisley Dodds and Lori Hinnant, 'Leaked Isis Documents Reveal Recruits Have Poor Grasp of Islamic Faith', *Independent* (16 August 2016).

18 Quoted in ibid.

19 J. Kiley Hamlin, Karen Wynn and Paul Bloom, 'Social Evaluation by Preverbal Infants', *Nature* (22 November 2007).

20 Paul Bloom, *Just Babies. The Origins of Good and Evil* (New York, 2013), p. 28.

21 J. Kiley Hamlin et al., 'Not Like Me = Bad: Infants Prefer Those Who Harm Dissimilar Others', *Psychological Science*, Vol. 24, Issue 4 (2013).

22 Karen Wynn said this on the CNN show *Anderson Cooper 360* on 15 February 2014.

23 Bloom, *Just Babies*, pp. 104–5.

24 The first meta-analysis, which included twenty-six studies, concluded that babies' preference for good guys is 'a well-established empirical finding'. But not everyone is convinced. Some scientists who repeated Hamlin's experiment saw the same effect, but others found no significant correlation. See Francesco Margoni and Luca Surian, 'Infants' Evaluation of Prosocial and Antisocial Agents: A Meta-Analysis', *Developmental Psychology*, Vol. 54, Issue 8 (2018).

25 Susan Seligson, 'Felix Warneken Is Overturning Assumptions about the Nature of Altruism', *Radcliffe Magazine* (Winter 2015).

26 In Warneken's TEDx Talk (titled: 'Need Help? Ask a 2-Year-Old'), available on YouTube, you can see a touching video of a child climbing out of a ball pit to help someone in need.

27 Not only that, if you do reward a toddler with candy or a toy, Warneken found they subsequently help *less*, since that was not their motive (see Chapter 13 on intrinsic motivation). Felix Warneken and Michael Tomasello, 'Extrinsic Rewards Undermine Altruistic Tendencies in 20-Month-Olds', *Development Psychology*, Vol. 44, Issue 6 (2008).

28 Stephen G. Bloom, 'Lesson of a Lifetime', *Smithsonian Magazine* (September 2005).

29 Quoted in ibid.

30 Quoted in ibid.

31 Rebecca S. Bigler and Meagan M. Patterson, 'Social Stereotyping and Prejudice in Children. Insights for Novel Group Studies', in Adam Rutland, Drew Nesdale and Christia Spears Brown (eds), *The Wiley Handbook of Group Processes in Children and Adolescents* (Oxford, 2017), pp. 184–202.

32 Yarrow Dunham, Andrew Scott Barron and Susan Carey, 'Consequences of "Minimal" Group Affiliations in Children', *Child Development*, Vol. 82, Issue 3 (2011), p. 808.

33 See also Hejing Zhang et al., 'Oxytocin Promotes Coordinated Out-group Attack During Intergroup Conflict in Humans', *eLife* (25 January 2019).

34 Apparently, I'm not alone. See Elijah Wolfson, 'Why We Cry on Planes', *The Atlantic* (1 October 2013).

35 Paul Bloom, *Against Empathy. The Case for Rational Compassion* (New York, 2016), p. 15.

36 Daniel Batson, 'Immorality from Empathy-induced Altruism: When Compassion and Justice Conflict,' *Journal of Personality and Social Psychology*, Vol. 68, Issue 6 (1995).

37 Michael N. Stagnaro and Paul Bloom, 'The Paradoxical Effect of Empathy on the Willingness to Punish', Yale University, unpublished manuscript (2016). See also Bloom, *Against Empathy*, p. 195.

38 Psychologists refer to this as the 'moralisation gap' – the tendency to perceive harm inflicted on us (or on those we care about) as somehow being much worse than any harm we inflict on others. An attack on a loved one so upsets us that we seek retribution, which we deem to be proportionate and justified when we do it, but completely excessive when the *other* does it, prompting us to strike back again. (You may have experienced this kind of escalating quarrel in relationships. The moralisation gap can also help us to understand

the decades of bloodshed in Israel and Palestine. Many people blame a lack of empathy, but I've come to believe there is rather too much empathy at work in the Middle East.)

39   George Orwell, 'Looking Back on the Spanish War' (August 1942).

40   Grossman, *On Killing*, p. 122.

41   Quoted in ibid., p. 126.

42   John Ellis, *The World War II Databook. The Essential Facts and Figures for All the Combatants* (London, 1993), Table 57, p. 257.

43   What, then, about the Rwandan genocide in 1994, when an estimated eight hundred thousand Tutsis and moderate Hutus were slaughtered? In the West, this example is often used to paint humans as bloodthirsty 'monsters', but that is largely because we know so little of the history. More recently, a historian has written, 'There is now ample evidence that the mass extermination of Rwandan citizens was the culmination of a carefully prepared, well-organized, bureaucratic campaign, using modern means of mass communication, propaganda, civil administration, and military logistics'. The actual murders were carried out by a small minority, in which an estimated 97 per cent of Hutus did not take part. See Abram de Swaan, *The Killing Compartments. The Mentality of Mass Murder* (New Haven and London, 2015), p. 90.

44   Łukasz Kamieński, *Shooting Up. A Short History of Drugs and War* (Oxford, 2016).

45   Lee, *Up Close and Personal*, p. 27.

46   Snipers much more often belong to the 1 to 2 per cent of soldiers who are psychopaths and have no natural aversion to killing. See Susan Neiman, *Moral Clarity. A Guide for Grown-Up Idealists* (Princeton, 2008), p. 372.

47   Dave Grossman, 'Hope on the Battlefield', in Dacher Keltner, Jason Marsh and Jeremy Adam Smith (eds), *The Compassionate Instinct. The Science of Human Goodness* (New York, 2010), p. 41.

48   Grossman, *On Killing*, p. 178.

49   Many soldiers who fought in the First and Second World Wars were also traumatised; however, Vietnam was comparatively much more traumatic. Of course, other factors were also to blame (such as the chilly reception Vietnam vets received on their return), but all the evidence suggests that the biggest was how the soldiers were conditioned to kill. Three recent studies among 1,200 veterans of Vietnam, 2,797 of Iraq and 317 of the Gulf War have shown

that soldiers who killed (enabled by their conditioning) are at a substantially higher risk of PTSD. See Shira Maguen et al., 'The Impact of Reported Direct and Indirect Killing on Mental Health Symptoms in Iraq War Veterans', *Journal of Traumatic Stress*, Vol. 23, Issue 1 (2010); Shira Maguen et al., 'The impact of killing on mental health symptoms in Gulf War veterans', *Psychological Trauma. Theory, Research, Practice, and Policy*, Vol. 3, Issue 1 (2011); and Shira Maguen et al., 'The Impact of Killing in War on Mental Health Symptoms and Related Functioning', *Journal of Traumatic Stress*, Vol. 45, Issue 10 (2009).

50 Frederick L. Coolidge, Felicia L. Davis and Daniel L. Segal, 'Understanding Madmen: A DSM-IV Assessment of Adolf Hitler', *Individual Differences Research*, Vol. 5, Issue 1 (2007).

51 Bond, *The Power of Others*, pp. 94–5.

11 How Power Corrupts

1 Quoted in Miles J. Unger, *Machiavelli. A Biography* (London, 2011), p. 8.

2 Niccolò Machiavelli, *The Prince*, translated by James B. Atkinson (Cambridge, Mass., 2008), p. 271. Originally published in 1532.

3 Machiavelli, *The Discourses*. Quoted in ibid., p. 280.

4 Dacher Keltner, *The Power Paradox. How We Gain and Lose Influence* (New York, 2017), pp. 41–9.

5 Melissa Dahl, 'Powerful People Are Messier Eaters, Maybe', *The Cut* (13 January 2015).

6 See for an overview: Aleksandra Cislak et al., 'Power Corrupts, but Control Does Not: What Stands Behind the Effects of Holding High Positions', *Personality and Social Psychology Bulletin*, Vol. 44, Issue 6 (2018), p. 945.

7 Paul K. Piff et al., 'Higher Social Class Predicts Increased Unethical Behaviour', *Proceedings of the National Academy of Sciences*, Vol. 109, Issue 11 (2012), pp. 4086–91.

8 Benjamin Preston, 'The Rich Drive Differently, a Study Suggests', *New York Times* (12 August 2013).

9 See Jeremy K. Boyd, Katherine Huynh and Bonnie Tong, 'Do wealthier drivers cut more at all-way stop intersections? Mechanisms underlying the relationship between social class and unethical

behavior' (University of California, San Diego, 2013). And Beth Morling et al., 'Car Status and Stopping for Pedestrians (#192)', *Psych File Drawer* (2 June 2014).

10 Keltner, *The Power Paradox*, pp. 99–136.

11 Jeremy Hogeveen, Michael Inzlicht and Suhkvinder S. Obhi, 'Power Changes How the Brain Responds to Others', *Journal of Experimental Psychology*, Vol. 143, Issue 2 (2014).

12 Jerry Useem, 'Power Causes Brain Damage', *The Atlantic* (July/August 2017).

13 See, for example, M. Ena Inesi et al., 'How Power Corrupts Relationships: Cynical Attributions for Others' Generous Acts', *Journal of Experimental Social Psychology*, Vol. 48, Issue 4 (2012), pp. 795–803.

14 Keltner, *The Power Paradox*, pp. 137–58.

15 Varun Warrier et al., 'Genome-Wide Analyses of Self-Reported Empathy: Correlations with Autism, Schizophrenia, and Anorexia Nervosa', *Nature, Translational Psychiatry* (12 March 2018).

16 Lord Acton, 'Letter to Bishop Mandell Creighton' (5 April 1887), published in J. N. Figgis and R. V. Laurence (eds), *Historical Essays and Studies* (London, 1907).

17 Frans de Waal, *Chimpanzee Politics. Power and Sex Among Apes* (Baltimore, 2007), p. 4. Originally published in 1982.

18 Frans de Waal and Frans Lanting, *Bonobo. The Forgotten Ape* (Berkeley, 1997).

19 Natalie Angier, 'In the Bonobo World, Female Camaraderie Prevails', *New York Times* (10 September 2016).

20 Frans de Waal, 'Sex as an Alternative to Aggression in the Bonobo', in Paul R. Abramson and Steven D. Pinkerton, *Sexual Nature/Sexual Culture* (Chicago, 1995), p. 37.

21 Christopher Boehm, 'Egalitarian Behavior and Reverse Dominance Hierarchy', *Current Anthropology*, Vol. 34, Issue 3 (1993), p. 233.

22 Christina Starmans, Mark Sheskin and Paul Bloom, 'Why People Prefer Unequal Societies', *Nature Human Behaviour*, Vol. 1, Issue 4 (2017).

23 See also Rutger Bregman and Jesse Frederik, 'Waarom vuilnismannen meer verdienen dan bankiers', *De Correspondent* (2015).

24 The best-known advocate of this theory is Yuval Noah Harari, in his book *Sapiens* (2011).

25 Robin Dunbar, *How Many Friends Does One Person Need? Dunbar's Number and Other Evolutionary Clues* (Cambridge, Mass., and London, 2010), p. 26.

26 The most persuasive defence of this theory is in Ara Norenzayan, *Big Gods* (2013). See also Harvey Whitehouse et al., 'Complex Societies Precede Moralizing Gods Throughout World History', *Nature* (20 March 2019) and Edward Slingerland et al., 'Historians Respond to Whitehouse et al. (2019), "Complex Societies Precede Moralizing Gods Throughout World History"', *PsyArXiv Preprints* (2 May 2019).

27 Harari, *Sapiens*, p. 34.

28 Douglas W. Bird et al., 'Variability in the organization and size of hunter-gatherer groups: Foragers do not live in small-scale societies', *Journal of Human Evolution* (June 2019).

29 Hill et al., 'Hunter-Gatherer Inter-Band Interaction Rates. Implications for Cumulative Culture'.

30 Graeber and Wengrow, 'How to Change the Course of Human History (at Least, the Part That's Already Happened)'.

31 Machiavelli, *The Prince*, p. 149.

32 David Graeber, *The Utopia of Rules. On Technology, Stupidity and the Secret Joys of Bureaucracy* (Brooklyn and London, 2015), pp. 31–3.

33 This is why serious economists were able to predict early on that the myth we call 'Bitcoin' was doomed to fail while the dollar will prevail for many more decades. The dollar is backed by the world's most powerful army, whereas the Bitcoin is backed only by belief.

34 Harari, *Sapiens*, p. 153.

35 Quoted in Noam Chomsky, 'What is the Common Good?', *Truthout* (7 January 2014).

36 Just how effective shaming can be was recently proved yet again by the #MeToo movement. Starting in October 2017, thousands of women took down a succession of male aggressors in a way strongly reminiscent of how bonobo females rein in antagonists and nomadic tribes tame bullies. By publicly humiliating the perpetrators, others will think twice before engaging in similar behavior.

37 Olivia Solon, 'Crazy at the Wheel: Psychopathic CEOs are Rife in Silicon Valley, Experts Say', *Guardian* (15 March 2017). See also Karen Landay, Peter, D. Harms and Marcus Credé, 'Shall We Serve the Dark Lords? A Meta-Analytic Review of Psychopathy and Leadership', *Journal of Applied Psychology* (August 2018).

12 What the Enlightenment Got Wrong

1   C. P. Snow, 'Science and Government', The Godkin Lectures (1960).
2   David Hume, 'Of the Independency of Parliament', in Essays,
    Moral, Political, and Literary (1758, Part 1).
3   See the famous poem by Bernard Mandeville 'The Grumbling
    Hive: Or, Knaves turn'd Honest', *The Fable of The Bees: or, Private
    Vices, Public Benefits* (1714).
4   Marshall Sahlins, *The Western Illusion of Human Nature* (Chicago,
    2008), pp. 72–6.
5   His Holiness Pope Francis, 'Why the Only Future Worth Building
    Includes Everyone', TED Talks (April 2017).
6   Ara Norenzayan, *Big Gods* (Princeton, 2013), p. 75.
7   If you don't believe it, this will set you straight: Hans Rosling,
    *Factfulness. Ten Reasons We're Wrong About the World – and Why
    Things Are Better Than You Think* (New York, 2018).
8   For an overview, see the first chapter of my previous book *Utopia for
    Realists* (London, 2017).
9   See, for example, Zygmunt Bauman, *Modernity and the Holocaust*
    (Ithaca, 1989), and Roger Griffin, *Modernism and Fascism. The
    Sense of a Beginning under Mussolini and Hitler* (Basingstoke,
    2007).

Part 4   A New Realism

1   Quoted in Hanna Rosin and Alix Spiegel, 'How to Become Batman',
    NPR (23 January 2015).
2   Quoted in Katherine Ellison, 'Being Honest About the Pygmalion
    Effect', *Discover Magazine* (December 2015).
3   Ibid.
4   Dov Eden, 'Self-Fulfilling Prophecy and the Pygmalion Effect in
    Management', *Oxford Bibliographies* (20 October 2016).
5   Lee Jussim and Kent D. Harber, 'Teacher Expectations and Self-
    Fulfilling Prophecies: Knowns and Unknowns, Resolved and
    Unresolved Controversies', *Personality and Social Psychology
    Review* (1 May 2005). See also Rhona S. Weinstein, 'Pygmalion at
    50: harnessing its power and application in schooling', *Educational
    Research and Evaluation* (11 December 2018).

6   Dov Eden, quoted in Ellison, 'Being Honest About the Pygmalion Effect'.

7   Franklin H. Silverman, 'The "Monster" Study', *Journal of Fluency Disorders*, Vol. 13, Issue 3 (1988).

8   John C. Edwards, William McKinley and Gyewan Moon, 'The enactment of organizational decline: The self-fulfilling prophecy', *International Journal of Organizational Analysis*, Vol. 10, Issue 1 (2002).

9   Daisy Yuhas, 'Mirror Neurons Can Reflect Hatred', *Scientific American* (1 March 2013).

10   John Maynard Keynes, *The General Theory of Employment, Interest, and Money* (London, 1936), Chapter 12.

11   Dan Ariely, 'Pluralistic Ignorance', *YouTube* (16 February 2011).

12   Pinker, *The Better Angels of Our Nature* (2011), pp. 561–5.

13 The Power of Intrinsic Motivation

1   Hedwig Wiebes, 'Jos de Blok (Buurtzorg): "Ik neem nooit zomaar een dag vrij",' *Intermediair* (21 October 2015).

2   Ibid.

3   Ibid.

4   Haico Meijerink, 'Buurtzorg: "Wij doen niet aan strategische flauwekul",' *Management Scope* (8 October 2014).

5   Gardiner Morse, 'Why We Misread Motives', *Harvard Business Review* (January 2003).

6   Quoted in ibid.

7   Frederick Taylor, *The Principles of Scientific Management* (New York, 1911), Chapter 2, p. 59.

8   Quoted in Robert Kanigel, *The One Best Way. Frederick Winslow Taylor and the Enigma of Efficiency* (Cambridge, 2005), p. 499.

9   Edward L. Deci, 'Effects of Externally Mediated Rewards on Intrinsic Motivation', *Journal of Personality and Social Psychology*, Vol. 1, Issue 1 (1971), p. 114.

10   Quoted in Karen McCally, 'Self-Determined', *Rochester Review* (July–August 2010).

11   Uri Gneezy and Aldo Rustichini, 'A Fine is a Price', *Journal of Legal Studies*, Vol. 29, Issue 1 (2000).

12   Samuel Bowles and Sandra Polanía Reyes, 'Economic Incentives and Social Preferences: A Preference-Based Lucas Critique of Public Policy', *University of Massachusetts Amherst Working Papers* (2009).

13 Amit Katwala, 'Dan Ariely: Bonuses boost activity, not quality', *Wired* (February 2010).

14 *Perceptions Matter: The Common Cause UK Values Survey*, Common Cause Foundation (2016).

15 Milton Friedman, 'The Methodology of Positive Economics', in *Essays in Positive Economics* (Chicago, 1966).

16 Sanford E. DeVoe and Jeffrey Pfeffer, 'The Stingy Hour: How Accounting for Time Affects Volunteering', *Personality and Social Psychology Bulletin*, Vol. 36, Issue 4 (2010).

17 Steve Crabtee, 'Worldwide, 13% of Employees Are Engaged at Work', *Gallup* (8 October 2013).

18 Wiljan van den Berge and Bas ter Weel, *Baanpolarisatie in Nederland. CPB Policy Brief*, Statistics Netherlands (2015), p. 14.

19 Quoted in Enzo van Steenbergen and Jeroen Wester, 'Hogepriester van de kleinschalige zorg', *NRC Handelsblad* (12 March 2016). Some competitors have criticised the fact that Buurtzorg shifts patients with severe problems onto other care suppliers, but there's no evidence to back this up. On the contrary, KPMG consultant David Ikkersheim's research revealed that Buurtzorg was better and cheaper even after correcting for care load. See David Ikkersheim, 'Buurtzorg: hoe zat het ook alweer?,' *Skipr* (9 May 2016).

20 Quoted in Stevo Akkerman, 'Betere zorg zonder strategische fratsen', *Trouw* (1 March 2016).

21 Quoted in The Corporate Rebels, 'FAVI. How Zobrist Broke Down Favi's Command-And-Control Structures', *corporate-rebels.com* (4 January 2017).

22 Patrick Gilbert, Nathalie Raulet Crozet and Anne-Charlotte Teglborg, 'Work Organisation and Innovation – Case Study: FAVI, France', *European Foundation for the Improvement of Living and Working Conditions* (2013).

14 Homo ludens

1 Stephen Moss, *Natural Childhood Report* (National Trust), p. 5.

2 John Bingham, 'British Children among Most Housebound in World', *Daily Telegraph* (22 March 2016).

3 S. L. Hofferth and J. F. Sandberg, 'Changes in American Children's Time, 1981–1997', in S. L. Hofferth and J. Owens (eds), *Children at*

the Millennium: Where Have We Come from? Where Are We Going?
(Stamford, 2001).

4   Peter Gray, 'The Decline of Play and the Rise of Psychopathology in
Children and Adolescents', *American Journal of Play*, Vol. 23, Issue 4
(2011), p. 450.

5   Jantje Beton/Kantar Public (TNS NIPO), *Buitenspelen Onderzoek
2018*, jantjebeton.nl (17 April 2018)

6   Frank Huiskamp, 'Rapport: Nederlandse leerlingen zijn niet
gemotiveerd', *NRC Handelsblad* (16 April 2014).

7   *Gezinsrapport. Een portret van het gezinsleven in Nederland*,
Netherlands Institute for Social Research (The Hague, 2011).

8   Rebecca Rosen, 'America's Workers: Stressed Out, Overwhelmed,
Totally Exhausted', *The Atlantic* (27 March 2014).

9   Jessica Lahey, 'Why Kids Care More About Achievement Than
Helping Others', *The Atlantic* (25 June 2014).

10 See, for example, C. Page Moreau and Marit Gundersen
Engeset, 'The Downstream Consequences of Problem-Solving
Mindsets: How Playing with LEGO Influences Creativity', *Journal of
Marketing Research*, Vol. 53, Issue 1 (2016).

11 Peter Gray, 'The Play Deficit', *Aeon* (18 September 2013).

12 *How to Tame a Fox (And Build a Dog)* (2017), p. 73.

13 Sarah Zielinski, 'Five Surprising Animals That Play', *ScienceNews*
(20 February 2015).

14 Johan Huizinga, *Homo Ludens. Proeve eener bepaling van het spel-
element der cultuur* (1938).

15 Peter Gray, 'Play as a Foundation for Hunter Gatherer Social
Existence', *American Journal of Play* (Spring 2009).

16 Jared Diamond, *The World Until Yesterday. What Can We Learn From
Traditional Societies?* (London, 2013), p. 204.

17 Ibid., p. 194.

18 Quoted in J. Mulhern, *A History of Education, a Social Interpretation*
(New York, 1959), p. 383.

19 James C. Scott, *Two Cheers for Anarchism. Six Easy Pieces on
Autonomy, Dignity and Meaningful Work and Play* (Princeton, 2012),
pp. 54–5.

20 The seminal work on this process is Eugen Weber, *Peasants into
Frenchmen: The Modernization of Rural France, 1870–1914* (Stanford,
1976).

21 Howard P. Chudacoff, *Children at Play. An American History*
(New York, 2008).

22 Peter Gray, 'The Decline of Play and the Rise of Psychopathology in Children and Adolescents' (2011).

23 Quoted in Robert Dighton, 'The Context and Background of the First Adventure Playground', *adventureplay.org.uk*

24 Quoted in Colin Ward, *Anarchy in Action* (London, 1996), p. 89.

25 Quoted in Arvid Bengtsson, *Adventure Playgrounds*, Crosby Lockwood (1972), pp. 20–21.

26 Quoted in Penny Wilson, 'children are more complicated than kettles. the life and work of Lady Allen of Hurtwood', *theinternationale.com* (2013).

27 Ibid.

28 Ibid.

29 Mariana Brussoni et al., 'What is the Relationship between Risky Outdoor Play and Health in Children? A Systematic Review', *International Journal of Environmental Research and Public Health*, Vol. 12, Issue 6 (8 June 2015).

30 Quoted in Rebecca Mead, 'State of Play', *The New Yorker* (5 July 2010).

31 Erving Goffman, 'On the Characteristics of Total Institutions' (1957).

32 Robin Bonifas, *Bullying Among Older Adults. How to Recognize and Address an Unseen Epidemic* (Baltimore, 2016).

33 Matt Sedensky, 'A surprising bullying battleground: Senior centers', Associated Press (13 May 2018).

34 Randall Collins, *Violence. A Micro-sociological Theory* (Princeton, 2008), p. 166.

35 Take Hogwarts, the school of Harry Potter fame. In J. K. Rowling's captivating fantasy world it's a magical place, but in reality I suspect it would be hell for many children. Kids are grouped by age (in classes) and by personality (in houses, e.g. Gryffindor and Slytherin). Authority figures encourage competition with a complex point system. If you want to leave, your only options are Christmas and summer break. Educationalists agree that Hogwarts is a recipe for a bullying culture.

36 Don't get me wrong, there are basic skills, like reading and writing, that people living in modern society can't do without. And there are kids with less aptitude to learn these skills. In such cases, expert instruction by a trained teacher is essential.

37 Robert Dur and Max van Lent, 'Socially Useless Jobs', Tinbergen Institute Discussion Paper (2 May 2018).

38 David Graeber, 'On the Phenomenon of Bullshit Jobs: A Work Rant', *Strike! Magazine* (August 2013).

39 Ivan Illich, *Deschooling Society* (New York, 1971).

40 Peter Gray, *Free to Learn. Why Unleashing the Instinct to Play Will Make Our Children Happier, More Self-Reliant, and Better Students for Life* (New York, 2013).

41 Quoted in Lois Holzman, 'What's the Opposite of Play?', *Psychology Today* (5 April 2016).

42 '"Depression: Let's Talk" Says WHO, As Depression Tops List of Causes of Ill Health', World Health Organization (30 March 2017).

43 Peter Gray, 'Self-Directed Education – Unschooling and Democratic Schooling', *Oxford Research Encyclopedia of Education* (April 2017).

15 This Is What Democracy Looks Like

1 Municipalities in Venezuela are a little like counties in some US states. But in Venezuela they're also where local government operates and mayors are elected.

2 Gabriel Hetland, 'Emergent Socialist Hegemony in Bolivarian Venezuela: The Role of the Party', in Susan J. Spronk and Jeffery R. Webber, *Crisis and Contradiction: Marxist Perspectives on Latin America in the Global Political Economy* (Leiden, 2015), p. 131.

3 Gabriel Hetland, 'How to Change the World: Institutions and Movements Both Matter', *Berkeley Journal of Sociology* (3 November 2014).

4 For a cogent account, see Gabriel Hetland, 'Grassroots Democracy in Venezuela', *The Nation* (30 January 2012).

5 Quoted in ibid.

6 Dmytro Khutkyy, 'Participatory budgeting: An empowering democratic institution', *Eurozine* (31 October 2017).

7 *Brazil: Toward a More Inclusive and Effective Participatory Budget in Porto Alegre* (World Bank, 2008), p. 2.

8 Quoted in Martin Calisto Friant, 'Sustainability From Below: Participatory Budgeting in Porto Alegre', First Ecuadorian Congress of Urban Studies (November 2017), p. 13.

9 Paolo Spada, 'The Economic and Political Effects of Participatory Budgeting', Congress of the Latin American Studies Association (2009).

10   Esteban Ortiz-Ospina and Max Roser, 'Trust', *OurWorldInData.org* (2018).

11   For a critique of this thesis, see Omar Encarnación, *The Myth of Civil Society. Social Capital and Democratic Consolidation in Spain and Brazil* (Basingstoke, 2003).

12   Quoted in 'Porto Alegre's Budget Of, By, and For the People', *Yes! Magazine* (31 December 2002).

13   Ginia Bellafante, 'Participatory Budgeting Opens Up Voting to the Disenfranchised and Denied', *New York Times* (17 April 2015).

14   Mona Serageldin et al., 'Assessment of Participatory Budgeting in Brazil', Harvard University Center for Urban Development Studies (2005), p. 4.

15   Gianpaolo Baiocchi, 'Participation, Activism, and Politics: The Porto Alegre Experiment in Deliberative Democratic Theory', in Archon Fung and Erik Olin Wright (eds), *Deepening Democracy. Institutional Innovations in Empowered Participatory Governance* (New York, 2001), p. 64.

16   Alana Semuels, 'The City That Gave Its Residents $3 Million', *The Atlantic* (6 November 2014).

17   Baiocchi, 'Participation, Activism, and Politics: The Porto Alegre Experiment in Deliberative Democratic Theory'.

18   Gianpaolo Baiocchi and Ernesto Ganuza, 'Participatory Budgeting as if Emancipation Mattered', *Politics & Society*, Vol. 42, Issue 1 (2014), p. 45.

19   George Monbiot, *Out of the Wreckage. A New Politics for an Age of Crisis* (London, 2017), p. 130.

20   Anne Pordes Bowers and Laura Bunt, 'Your Local Budget. Unlocking the Potential of Participatory Budgeting', *Nesta* (2010).

21   Gianpaolo Baiocchi, 'Participation, Activism, and Politics: The Porto Alegre Experiment and Deliberative Democratic Theory', *Politics & Society*, Vol. 29, Issue 1 (2001), p. 58.

22   World Bank researchers also concluded the rapid advances owed everything to participatory budgeting. The share of the city budget allocated to healthcare and education increased from 13 per cent in 1985 to 40 per cent in 1996. See Serageldin et al., 'Assessment of Participatory Budgeting in Brazil'.

23   Patrick Kingsley, 'Participatory democracy in Porto Alegre', *Guardian* (10 September 2012).

24   Serageldin et al., 'Assessment of Participatory Budgeting in Brazil'.

25 Michael Touchton and Brian Wampler, 'Improving Social Well-Being Through New Democratic Institutions', *Comparative Political Studies*, Vol. 47, Issue 10 (2013).

26 'Back to the Polis: Direct Democracy', *The Economist* (17 September 1994).

27 David Van Reybrouck, *Against Elections. The Case for Democracy* (London, 2016).

28 'Communism', oxforddictionaries.com.

29 Graeber, *Debt*, pp. 94–102.

30 Garrett Hardin, 'The Tragedy of the Commons', *Science*, Vol. 162, Issue 3859 (13 December 1968).

31 John Noble Wilford, 'A Tough-minded Ecologist Comes to Defense of Malthus', *New York Times* (30 June 1987).

32 Ian Angus, 'The Myth of the Tragedy of the Commons', *Climate & Capitalism* (25 August 2008).

33 John A. Moore, 'Science as a Way of Knowing – Human Ecology', *American Zoologist*, Vol. 25, Issue 2 (1985), p. 602.

34 Tim Harford, 'Do You Believe in Sharing?' *Financial Times* (30 August 2013).

35 Ibid.

36 Officially: The Sveriges Riksbank Prize in Economic Sciences in Memory of Alfred Nobel.

37 Tine de Moor, 'The Silent Revolution: A New Perspective on the Emergence of Commons, Guilds, and Other Forms of Corporate Collective Action in Western Europe', *International Review of Social History*, Vol. 53, Issue S16 (December 2008).

38 The classic work on this process is Karl Polanyi, *The Great Transformation. The Political and Economic Origins of Our Time* (Boston, 2001). Originally published in 1944.

39 Tine de Moor, 'Homo Cooperans. Institutions for collective action and the compassionate society', Utrecht University Inaugural Lecture (30 August 2013).

40 See, for example, Paul Mason, *Postcapitalism. A Guide to Our Future* (London, 2015).

41 See, for example, Shoshana Zuboff, *The Age of Surveillance Capitalism. The Fight for a Human Future at the New Frontier of Power* (London, 2019).

42 Damon Jones and Ioana Elena Marinescu, 'The Labor Market Impacts of Universal and Permanent Cash Transfers: Evidence

from the Alaska Permanent Fund', *NBER Working Paper* (February 2018).

43 I've also written about this study in North Carolina and about universal basic income elsewhere. See *Utopia for Realists. And How We Can Get There* (London, 2017), pp. 51–4. I now prefer the term 'citizen's dividend' over 'basic income' to underscore that we're talking about proceeds from communal property.

44 Peter Barnes, *With Liberty and Dividends For All. How To Save Our Middle Class When Jobs Don't Pay Enough* (Oakland, 2014).

45 Scott Goldsmith, 'The Alaska Permanent Fund Dividend: An Experiment in Wealth Distribution', *Basic Income European Network* (September 2002), p. 7.

Part 5   The Other Cheek

1 Michael Garofalo, 'A Victim Treats His Mugger Right', NPR Story Corps (28 March 2008).

2 Matthew 5:46.

16   Drinking Tea with Terrorists

1 For a clear overview of Norway's prison system, see Ryan Berger, 'Kriminalomsorgen: A Look at the World's Most Humane Prison System in Norway', *SSRN* (11 December 2016).

2 A guard says this in Michael Moore's documentary, *Where to Invade Next?* (2015).

3 Quoted in Baz Dreizinger, 'Norway Proves That Treating Prison Inmates As Human Beings Actually Works', *Huffington Post* (8 March 2016).

4 'About the Norwegian Correctional Service', www.kriminalomsorgen.no (visited 17 December 2018).

5 Dreizinger, 'Norway Proves That Treating Prison Inmates As Human Beings Actually Works'.

6 Manudeep Bhuller et al., 'Incarceration, Recidivism, and Employment', Institute of Labor Economics (June 2018).

7 Berger 'Kriminalomsorgen: A Look at the World's Most Humane Prison System in Norway', p. 20.

8 Erwin James, 'Bastoy: the Norwegian Prison That Works', *Guardian* (4 September 2013).

9 Genevieve Blatt et al., *The Challenge of Crime in a Free Society*, President's Commission on Law Enforcement and Administration of Justice (1967), p. 159.

10 Ibid., p. 173.

11 Jessica Benko, 'The Radical Humaneness of Norway's Halden Prison', *New York Times* (26 March 2015).

12 Robert Martinson, 'What Works? Questions and Answers about Prison Reform', *The Public Interest* (Spring 1974).

13 Michelle Brown, *The Culture of Punishment: Prison, Society, and Spectacle* (New York, 2009), p. 171.

14 Robert Martinson, 'New Findings, New Views: A Note of Caution Regarding Sentencing Reform', *Hofstra Law Review*, Vol. 7, Issue 2 (1979).

15 Quoted in Adam Humphreys, 'Robert Martinson and the Tragedy of the American Prison', *Ribbonfarm* (15 December 2016).

16 Quoted in Jerome G. Miller, 'The Debate on Rehabilitating Criminals: Is It True that Nothing Works?' *Washington Post* (March 1989).

17 Richard Bernstein, 'A Thinker Attuned to Thinking; James Q. Wilson Has Insights, Like Those on Cutting Crime, That Tend To Prove Out', *New York Times* (22 August 1998).

18 'James Q. Wilson Obituary', *The Economist* (10 March 2012).

19 James Q. Wilson, *Thinking About Crime* (New York, 1975), pp. 172–3.

20 Quoted in Timothy Crimmins, 'Incarceration as Incapacitation: An Intellectual History', *American Affairs*, Vol. II, Issue 3 (2018).

21 George L. Kelling and James Q. Wilson, 'Broken Windows', *The Atlantic* (March 1982).

22 Gladwell, *The Tipping Point*, p. 141.

23 Ibid., p. 142.

24 Ibid., p. 143.

25 Holman W. Jenkins, Jr, 'The Man Who Defined Deviancy Up', *The Wall Street Journal* (12 March 2011).

26 James Q. Wilson, 'Lock 'Em Up and Other Thoughts on Crime', *New York Times* (9 March 1975).

27 Gladwell, *The Tipping Point*, p. 145.

28 Quoted in ibid., p. 146.

29 'New York Crime Rates 1960–2016', disastercenter.com

30 Donna Ladd, 'Inside William Bratton's NYPD: Broken Windows Policing is Here to Stay', *Guardian* (8 June 2015).

31 Quoted in Jeremy Rozansky and Josh Lerner, 'The Political Science of James Q. Wilson', *The New Atlantis* (Spring 2012).

32 See Rutger Bregman, *Met de kennis van toen. Actuele problemen in het licht van de geschiedenis* (Amsterdam, 2012), pp. 238–45.

33 Anthony A. Braga, Brandon C. Welsh and Cory Schnell, 'Can Policing Disorder Reduce Crime? A Systematic Review and Meta-Analysis', *Journal of Research in Crime and Delinquency*, Vol. 52, Issue 4 (2015).

34 John Eterno and Eli Silverman, 'Enough Broken Windows Policing. We Need a Community-Oriented Approach', *Guardian* (29 June 2015).

35 P. J. Vogt, '#127 The Crime Machine', *Reply All* (podcast by Gimlet Media, 11 October 2018).

36 Dara Lind, 'Why You Shouldn't Take Any Crime Stats Seriously', *Vox* (24 August 2014). See also Liberty Vittert, 'Why the US Needs Better Crime Reporting Statistics', *The Conversation* (12 October 2018).

37 Michelle Chen, 'Want to See How Biased Broken Windows Policing Is? Spend a Day in Court', *The Nation* (17 May 2018).

38 Order itself turns out to be a matter of perception. In 2004, researchers at the University of Chicago asked a series of study subjects how many 'broken windows' they saw in a white and a black neighbourhood. Subjects uniformly rated a neighbourhood with more African Americans as more disorderly, even when the prevalence of litter, graffiti and group loitering were the same as in the white neighbourhood. See Robert J. Sampson and Stephen W. Raudenbush, 'Seeing Disorder: Neighborhood Stigma and the Social Construction of "Broken Windows"', *Social Psychology Quarterly*, Vol. 67, Issue 4 (2004). The sad thing is that Wilson and Kelling already predicted this in 1982 article in *The Atlantic*, writing: 'How do we ensure that [...] skin color or national origin [...] will not also become the basis for distinguishing the undesirable from the desirable? How do we ensure, in short, that the police do not become the agents of neighborhood bigotry? We can offer no wholly satisfactory answer to this important question.'

39 See Braga, Welsh, and Schnell, 'Can Policing Disorder Reduce Crime? A Systematic Review and Meta-Analysis'.

40 Quoted in Sarah Childress, 'The Problem with "Broken Windows" Policing', *Frontline* (28 June 2016).
41 Vlad Tarko, *Elinor Ostrom. An Intellectual Biography* (Lanham, 2017), pp. 32–40.
42 Arthur A. Jones and Robin Wiseman, 'Community Policing in Europe. An Overview of Practices in Six Leading Countries', Los Angeles Community Policing (lacp.org).
43 Sara Miller Llana, 'Why Police Don't Pull Guns in Many Countries', *Christian Science Monitor* (28 June 2015).
44 Quoted in Childress, 'The Problem with "Broken Windows" Policing'.
45 Beatrice de Graaf, *Theater van de angst. De strijd tegen terrorisme in Nederland, Duitsland, Italië en Amerika* (Amsterdam, 2010).
46 Quoted in Quirine Eijkman, 'Interview met Beatrice de Graaf over haar boek', *Leiden University* (25 January 2010).
47 Quoted in Joyce Roodnat, '"Het moest wel leuk blijven"', *NRC Handelsblad* (6 April 2006).
48 Quoted in Jon Henley, 'How Do You Deradicalise Returning Isis Fighters?', *Guardian* (12 November 2014).
49 Quoted in Hanna Rosin, 'How A Danish Town Helped Young Muslims Turn Away From ISIS', *NPR Invisibilia* (15 June 2016).
50 Quoted in Richard Orange, '"Answer hatred with love": how Norway tried to cope with the horror of Anders Breivik', *Guardian* (15 April 2012).
51 Prison Policy Initiative, 'North Dakota Profile' (prisonpolicy.org, visited 17 December 2018).
52 Quoted in Dylan Matthews and Byrd Pinkerton, 'How to Make Prisons More Humane', *Vox* (podcast, 17 October 2018).
53 Dashka Slater, 'North Dakota's Norway Experiment', *Mother Jones* (July/August 2017).
54 National Research Council, *The Growth of Incarceration in the United States. Exploring Causes and Consequences* (Washington DC, 2014), p. 33.
55 Francis T. Cullen, Cheryl Lero Jonson and Daniel S. Nagin, 'Prisons Do Not Reduce Recidivism. The High Cost of Ignoring Science', *The Prison Journal*, Vol. 91, Issue 3 (2011). See also M. Keith Chen and Jesse M. Shapiro, 'Do Harsher Prison Conditions Reduce Recidivism? A Discontinuity-based Approach', *American Law and Economics Review*, Vol. 9, Issue 1 (2007).

56 'Louis Theroux Goes to the Miami Mega-Jail', *BBC News* (20 May 2011).
57 Quoted in Berger, 'Kriminalomsorgen: A Look at the World's Most Humane Prison System in Norway', p. 23.
58 Quoted in Slater, 'North Dakota's Norway Experiment'.
59 Cheryl Corley, 'North Dakota Prison Officials Think Outside The Box To Revamp Solitary Confinement', *NPR* (31 July 2018).
60 Ibid.
61 Quoted in Slater, 'North Dakota's Norway Experiment'.

17 The Best Remedy for Hate, Injustice and Prejudice

1 Quoted in John Battersby, 'Mandela to Factions: Throw Guns Into Sea', *Christian Science Monitor* (26 February 1990).
2 My main source for the story about Constand and Abraham is Dennis Cruywagen's wonderful book *Brothers in War and Peace. Constand and Abraham Viljoen and the Birth of the New South Africa* (Cape Town/Johannesburg, 2014).
3 Ibid., p. 57.
4 Ibid., p. 62.
5 Maritza Montero and Christopher C. Sonn (eds), *Psychology of Liberation. Theory and Applications* (Berlin, Heidelberg, 2009), p. 100.
6 Aldous Huxley, *The Perennial Philosophy* (New York, 1945), p. 81.
7 Alfred McClung Lee and Norman Daymond Humphrey, *Race Riot, Detroit 1943* (Hemel Hempstead, 1968), p. 130.
8 Gordon Allport, *The Nature of Prejudice* (Reading, 1979), p. 277. Originally published in 1954. Researchers asked American soldiers the following question: 'Some Army divisions have companies which include Negro and white platoons. How would you feel about it if your outfit was set up something like that?' The share that responded 'I would dislike it very much' was 62 per cent in strictly segregated units and 7 per cent among men in companies that included black platoons.
9 Ira N. Brophy, 'The Luxury of Anti-Negro Prejudice', *Public Opinion Quarterly*, Vol. 9, Issue 4 (1945).
10 Richard Evans, *Gordon Allport: The Man and His Ideas* (New York, 1970).

11 Gordon Allport, 'Autobiography', in Edwin Boring and Gardner Lindzey (eds), *History of Psychology in Autobiography* (New York, 1967), pp. 3–25.

12 John Carlin, *Invictus. Nelson Mandela and the Game that Made a Nation* (London, 2009), p. 122.

13 Quoted in ibid., p. 123.

14 Ibid., p. 124

15 Ibid., p. 135.

16 Cruywagen, *Brothers in War and Peace*, p. 143.

17 Quoted in ibid., p. 158.

18 Quoted in Simon Kuper, 'What Mandela Taught Us', *Financial Times* (5 December 2013).

19 Quoted in Cruywagen, *Brothers in War and Peace*, p. 162.

20 Quoted in Carlin, *Invictus*, p. 252.

21 To which Pettigrew responded: 'Sir, you have paid me the highest honor!' Quoted in Frances Cherry, 'Thomas F. Pettigrew: Building on the Scholar-Activist Tradition in Social Psychology', in Ulrich Wagner et al. (eds), *Improving Intergroup Relations: Building on the Legacy of Thomas F. Pettigrew* (Oxford, 2008), p. 16.

22 Thomas F. Pettigrew, 'Contact in South Africa', *Dialogue*, Vol. 21, Issue 2 (2006), pp. 8–9.

23 Thomas F. Pettigrew and Linda R. Tropp, 'A Meta-Analytic Test of Intergroup Contact Theory', *Journal of Personality and Social Psychology*, Vol. 90, Issue 5 (2006).

24 Sylvie Graf, Stefania Paolini and Mark Rubin, 'Negative intergroup contact is more influential, but positive intergroup contact is more common: Assessing contact prominence and contact prevalence in five Central European countries', *European Journal of Social Psychology*, Vol. 44, Issue 6 (2014).

25 Erica Chenoweth, 'The Origins of the NAVCO Data Project (or: How I Learned to Stop Worrying and Take Nonviolent Conflict Seriously)', *Rational Insurgent* (7 May 2014).

26 Erica Chenoweth and Maria J. Stephan, 'How The World is Proving Martin Luther King Right About Nonviolence', *Washington Post* (18 January 2016). See also Maria J. Stephan and Erica Chenoweth, 'Why Civil Resistance Works. The Strategic Logic of Nonviolent Conflict', *International Security*, Vol. 33, Issue 1 (2008), pp. 7–44.

27 Quoted in Penny Andersen et al., *At Home in the World. The Peace Corps Story* (Peace Corps, 1996), p. vi.

28 Carlin, *Invictus*, p. 84.

29   Ibid., p. 252.

30   Ibid.

31   Quoted in Thomas F. Pettigrew, 'Social Psychological Perspectives on Trump Supporters', *Journal of Social and Political Psychology*, Vol. 5, Issue 1 (2017).

32   Ibid.

33   Chris Lawton and Robert Ackrill, 'Hard Evidence: How Areas with Low Immigration Voted Mainly for Brexit', *The Conversation* (8 July 2016). See also Rose Meleady, Charles Seger and Marieke Vermue, 'Examining the Role of Positive and Negative Intergroup Contact and Anti-Immigrant Prejudice in Brexit', *British Journal of Social Psychology*, Vol. 56, Issue 4 (2017).

34   Michael Savelkoul et al., 'Anti-Muslim Attitudes in The Netherlands: Tests of Contradictory Hypotheses Derived from Ethnic Competition Theory and Intergroup Contact Theory', *European Sociological Review*, Vol. 27, Issue 6 (2011).

35   Jared Nai, 'People in More Racially Diverse Neighborhoods Are More Prosocial', *Journal of Personality and Social Psychology*, Vol. 114, Issue 4 (2018), pp. 497–515.

36   Miles Hewstone, 'Consequences of Diversity for Social Cohesion and Prejudice: The Missing Dimension of Intergroup Contact', *Journal of Social Issues*, Vol. 71, Issue 2 (2015).

37   Matthew Goodwin and Caitlin Milazzo, 'Taking Back Control? Investigating the Role of Immigration in the 2016 Vote for Brexit', *British Journal of Politics and International Relations*, Vol. 19, Issue 3 (2017).

38   Quoted in Diane Hoekstra, 'De felle tegenstanders van toen gaan het azc in Overvecht missen', *Algemeen Dagblad* (29 September 2018). See also Marjon Bolwijn, 'In Beverwaard was woede om azc het grootst, maar daar is niets meer van te zien: "We hebben elkaar gek gemaakt"', *De Volkskrant* (1 February 2018).

39   Mark Twain, *The Innocents Abroad, or The New Pilgrims' Progress* (1869).

40   Rupert Brown, James Vivian and Miles Hewstone, 'Changing Attitudes through Intergroup Contact: the Effects of Group Membership Salience', *European Journal of Social Psychology*, Vol. 29, Issue 5–6 (21 June 1999).

41   Gordon W. Allport, 'Prejudice in Modern Perspective', The Twelfth Hoernlé Memorial Lecture (17 July 1956).

18 When the Soldiers Came Out of the Trenches

1 This phrase was coined by the historian George F. Kennan in the introduction to his book *The Decline of Bismarck's European Order: Franco-Russian Relations 1875–1890* (Princeton, 1979).

2 Malcolm Brown and Shirley Seaton, *Christmas Truce. The Western Front December 1914* (London, 2014), p. 68. Originally published in 1984.

3 Ibid., p. 71.

4 Ibid., p. 73.

5 Ibid., pp. 76–7

6 Malcolm Brown, *Peace in No Man's Land* (BBC documentary from 1981).

7 Luke Harding, 'A Cry of: Waiter! And the Fighting Stopped', *Guardian* (1 November 2003).

8 Brown and Seaton, *Christmas Truce*, p. 111.

9 Ibid., p. 115.

10 Quoted in Simon Kuper, 'Soccer in the Trenches: Remembering the WWI Christmas Truce', espn.com (25 December 2014).

11 Modern historians point out that although the Germans certainly committed war crimes in 1914, they were heavily embellished in British propaganda. Just how disastrous the effects of such fake news can be would not become fully apparent until twenty-five years later. When in the Second World War reports began coming in of Germans committing the most horrifying atrocities on a massive scale, a sizeable share of people in the UK and US doubted their veracity. Given how the press had exaggerated matters during the First World War, it seemed to make sense to take the stories about gas chambers with a grain of salt. See Jo Fox, 'Atrocity propaganda', British Library (29 January 2014).

12 Brown and Seaton, *Christmas Truce*, p. 126.

13 Thomas Vinciguerra, 'The Truce of Christmas, 1914', *New York Times* (25 December 2005).

14 Quoted in TED Stories, 'Colombia: Advertising Creates Peace', YouTube (24 January 2018).

15 Ibid.

16 Tom Vanden Brook, 'Propaganda That Works: Christmas Decorations', *USA Today* (13 August 2013).

17 Lara Logan, 'How Unconventional Thinking Transformed a War-Torn Colombia', CBS News, *60 Minutes* (11 December 2016).

18 Quoted in TED Stories, 'Colombia: Advertising Creates Peace'.

19 Jose Miguel Sokoloff in an interview with the author on 9 November 2017.

20 The costs of Operation Christmas amounted to $301,100. River of Light came to $263,000 and Mother's Voice $546,000.

21 Even FARC believed this, since during the peace talks they demanded a stop to MullenLowe's propaganda. It was costing them too many members.

22 Sibylla Brodzinsky, ' "Welcome to Peace": Colombia's Farc Rebels Seal Historic Disarmament', *Guardian* (27 June 2017).

23 Quoted in Vinciguerra, 'The Truce of Christmas, 1914'.

24 Brown and Seaton, *Christmas Truce*, p. 198.

25 Ibid., p. 248.

26 Ibid., p. 238.

27 Stanley Weintraub, *Silent Night* (London, 2001), p. 172.

28 Tony Ashworth, *Trench Warfare 1914–1918. The Live and Let Live System* (London, 2000), p. 224. Originally published in 1980.

29 Ibid., p. 24.

30 Ibid., p. 143.

31 Erin E. Buckels, Paul D. Trapnell and Delroy L. Paulhus, 'Trolls Just Want to Have Fun', *Personality and Individual Difference*, Vol. 67 (September 2014).

32 Jose Miguel Sokoloff, 'How Christmas Lights Helped Guerillas Put Down Their Guns', TED (October 2014).

EPILOGUE

1 Detlef Fetchenhauer and David Dunning, 'Why So Cynical? Asymmetric Feedback Underlies Misguided Skepticism Regarding the Trustworthiness of Others', *Psychological Science*, Vol. 21, Issue 2 (8 January 2010).

2 There are a number of elegant studies demonstrating that approaching other people as though their intentions are good causes their demeanour to change. Psychologists call this 'virtue labelling'. For example, in 1975 American psychologist Richard Miller did a study among elementary school children where a randomly selected group was told they were 'tidy'. With a second group, the researchers did their best to get the children to be tidier, whereas a third group was left entirely alone. Result? The first group proved

by far the tidiest. See Christian B. Miller, 'Should You Tell Everyone They're Honest?', *Nautilus* (28 June 2018).

3   Maria Konnikova, *The Confidence Game. The Psychology of the Con and Why We Fall for It Every Time*, (Edinburgh, 2016).

4   Bloom, *Against Empathy*, p. 167.

5   Quoted in Dylan Matthews, 'Zero-sum Trump. What You Learn from Reading 12 of Donald Trump's Books', *Vox.com* (19 January 2017).

6   Marina Cantacuzino, *The Forgiveness Project. Stories for a Vengeful Age* (London, 2016).

7   Lewis B. Smedes, *Forgive and Forget. Healing the Hurts We Don't Deserve* (San Francisco, 1984).

8   Donald W. Pfaff, *The Neuroscience of Fair Play. Why We (Usually) Follow the Golden Rule*, Dana Press (2007).

9   George Bernard Shaw, *Maxims for Revolutionists* (1903).

10 Matthieu Ricard, *Altruism. The Power of Compassion to Change Yourself and the World* (New York, 2015), pp. 58–63.

11 Ibid., p. 62.

12 Daniel Goleman and Richard Davidson, *The Science of Meditation. How to Change Your Brain, Mind and Body* (London, 2018). But also see Miguel Farias and Catherine Wikholm, *The Buddha Pill. Can Meditation Change You?* (London, 2015).

13 Paul Bloom, 'Empathy for Trump voters? No, thanks. Understanding? Yes', *Vox.com* (23 February 2017).

14 Bloom, *Against Empathy*, pp. 213–41.

15 Jarl van der Ploeg, '"Ze zullen altijd die enorm verliefde bom geluk blijven"', *De Volkskrant* (21 July 2014).

16 'In memoriam: LvdG (1984–2014)', *Propria Cures* (19 July 2014).

17 See, for example, Chung Sup Park, 'Applying "Negativity Bias" to Twitter: Negative News on Twitter, Emotions, and Political Learning', *Journal of Information Technology & Politics*, Vol. 12, Issue 4 (2015).

18 Chris Weller, 'Silicon Valley Parents Are Raising Their Kids Tech-Free – And It Should Be a Red Flag', *Business Insider* (18 February 18, 2018).

19 Rebecca Solnit, *Hope in the Dark. Untold Histories, Wild Possibilities* (Chicago, 2016), p. 23.

20 Fabian Wichmann, '4 Ways To Turn The Neo-Nazi Agenda On Its Head', *Huffington Post* (25 August 2017).

21 Matthew 6:2–6

22   The French philosopher Alexis de Tocqueville remarked on this a
     century and a half ago. 'Americans ... enjoy explaining almost every
     act of their lives on the principle of self-interest,' he wrote. Having
     met plenty of helpful people in his travels, Tocqueville believed
     Americans were doing themselves a disservice. 'But the Americans,'
     the philosopher observed, 'are hardly prepared to admit that they do
     give way to emotions of this sort.' See Dale T. Miller, 'The Norm of
     Self-Interest', *American Psychologist*, Vol. 54, Issue 12 (1999).
23   Ibid., p. 1057.
24   Matthew 6:14–16
25   James H. Fowler and Nicholas A. Christakis, 'Cooperative Behavior
     Cascades in Human Social Networks', *PNAS*, Vol. 107, Issue 12
     (2010).
26   Quoted in University of California, San Diego, 'Acts of Kindness
     Spread Surprisingly Easily: Just a Few People Can Make a
     Difference', *ScienceDaily* (10 March 2010).
27   Jonathan Haidt, 'Elevation and the Positive Psychology of Morality',
     in C. L. M. Keyes and J. Haidt (eds), *Flourishing: Positive Psychology
     and the Life Well-Lived*, American Psychological Association (2003),
     pp. 275–89.
28   Quoted in Jonathan Haidt, 'Wired to Be Inspired', in Dacher Keltner,
     Jason Marsh and Jeremy Adam Smith (eds), *The Compassionate
     Instinct. The Science of Human Goodness* (New York, 2010), p. 90.

INDEX

A NOTE ON THE AUTHOR

Rutger Bregman is one of Europe's most prominent young historians. *Utopia for Realists* was a *Sunday Times* and *New York Times* bestseller and has been translated from the Dutch into thirty-two languages. He has twice been nominated for the prestigious European Press Prize for his work at *The Correspondent*, and his writing has also been featured in the *Washington Post* and the *Guardian*. His TED talk, 'Poverty isn't a lack of character; it's a lack of cash', has been viewed more than three million times. He was ranked number 10 in the *Big Issue*'s Top 100 Changemakers of 2020.

@rcbregman | rutgerbregman.com